Mariana de Carvalho
Paulo Cordeiro

Potential distribution of the richness of endemic Caatinga birds

Mariana de Carvalho
Paulo Cordeiro

Potential distribution of the richness of endemic Caatinga birds

Priority areas for conservation

ScienciaScripts

Imprint

Any brand names and product names mentioned in this book are subject to trademark, brand or patent protection and are trademarks or registered trademarks of their respective holders. The use of brand names, product names, common names, trade names, product descriptions etc. even without a particular marking in this work is in no way to be construed to mean that such names may be regarded as unrestricted in respect of trademark and brand protection legislation and could thus be used by anyone.

Cover image: www.ingimage.com

This book is a translation from the original published under ISBN 978-3-330-76278-7.

Publisher:
Sciencia Scripts
is a trademark of
Dodo Books Indian Ocean Ltd. and OmniScriptum S.R.L publishing group

120 High Road, East Finchley, London, N2 9ED, United Kingdom
Str. Armeneasca 28/1, office 1, Chisinau MD-2012, Republic of Moldova, Europe
Managing Directors: Ieva Konstantinova, Victoria Ursu
info@omniscriptum.com

Printed at: see last page
ISBN: 978-620-8-39136-2

SUMMARY

The diversity, richness and number of endemics of the Caatinga have long been underestimated, but it is now known that it is an extremely heterogeneous biome, more diverse than any other exposed to the same conditions. Despite this, the Caatinga is on the list of the most altered Brazilian biomes, emphasising the recurring need for further studies. Little is known about the distribution patterns of bird richness in this biome, so the aim of this study is to create potential distribution maps of endemic birds of the Caatinga along a gradient of richness, in order to search for distribution patterns and zones of greater suitability for the species. In addition, the degree to which these species are present within protected areas will be investigated. Twenty-nine species had their potential geographical distributions estimated, in which occurrence data was collected from the Scientific Collection of the National Museum of Rio de Janeiro, the Zoology Museum of São Paulo, data from the literature, the SpeciesLink network and the *WikiAves* platform. Four algorithms were used for the modelling - *Bio-Clim*, *Mahalanobis* Distance, SVM and *MaxEnt*, and to standardise the definition of the distribution limits for each species, the cut-off limit estimated by the models was used as the lowest commission index, resulting in conservative distribution maps. A PCA was carried out with 19 bi-climatic files and one altitude file available on *WorldClim*, obtaining nine variables with low collinearity, selected for modelling. To draw up the potential distribution maps, the models were intersected in *QuantumGis* 2.2. The results of the models were superimposed, generating a richness map. This showed two large regions of greater richness, one associated with the Araripe Plateau and the Borborema Plateau, the other with the Upper São Francisco region. It was noted that the regions of greatest richness correspond to areas of higher altitude, with an average of 450m to 850m. Evaluating the representativeness of the Conservation Units existing in the biome, it was observed that the areas of greatest richness lack protected areas, and are mainly represented by the surroundings of the Chapada do Araripe; eastern Paraíba on the border with Pernambuco; small portions of southern Bahia; extreme northern Minas Gerais; and eastern Rio Grande do Norte on the border with Paraíba, emphasising the need to implement Conservation Units in these regions. In future studies, it will be important to include a greater number of species, allowing for greater detail on the areas of richness in the Caatinga. In addition, it would be interesting to ascertain the similarity between Conservation Units and priority areas to be conserved, in order to create conservation strategies in environments that harbour poorly protected species. In general terms, it will only be possible to safeguard the biodiversity of the Caatinga through efforts supported by different studies that prove the uniqueness of evolutionary, biogeographical and natural history patterns.

Key words: Avifauna, Conservation, Potential distribution modelling, Richness patterns, Semi-arid.

Summary

CHAPTER 1

Introduction

From the 19th century onwards, the first direct contacts with the avifauna of certain regions of the Caatinga began, mainly attributed to the naturalist Prince Maximilian of Wied-Neuwied, followed by the naturalistic investigations of Johann Baptist von Spix and Karl Friedrich Phillipp von Martius in north-eastern Brazil (Pacheco 2004). There was also a visit by ornithologist William A. Forbes, who was able to make marginal contact with the avifauna of the Sertões, since his excursions were mainly concentrated in the Zona da Mata and Agreste (Pacheco 2004). Another highlight was the coleopterist Edmond Gounelle, who in the last two decades collected hummingbirds in the caatingas of the states of Pernambuco and Bahia (Pacheco 2004; Gounelle 1909 & Pinto 1940 apud Pacheco 2004). Numerous other initiatives from the 19th century included collecting birds in the Caatinga region, but most are poorly documented (Pacheco 2004). The 19th century saw most of the knowledge of northeastern birds concentrated in the Atlantic Rainforest, in the various specimens without mention of a specific locality or in Wied's results, which did not leave a clear picture of the avifauna of the Caatinga (Pacheco 2004).

It wasn't until the 20th century that there was a great advance in knowledge about the avifauna of the semi-arid region. This stage began with the contributions of Ichthyologist Franz Steindachner and Ornithologist Otmar Reiser (Hellmayr 1929). From then on, the number of expeditions and collections to the Caatinga grew and, consequently, knowledge about the local avifauna increased (Pacheco 2004).

However, even today the Caatinga lacks knowledge and is the least studied biome in the country (MMA 1999; 2002; Prado 2003). For a long time, the entire region covered by the biome was considered to be homogeneous, with low biodiversity, low rates of endemism and little change. In contrast to this perspective, recent studies point to this region as being rich in biodiversity, environments and endemism. It has the highest diversity records when compared to other biomes exposed to the same conditions (MMA 2002; 2007; Alves 2007; Alves et al. 2009; ICMBio 2011). In addition, it is recognised as an important area of endemism, including for birds (MMA 2007; ICMBio 2011).

The Caatinga is considered to be the second most altered ecosystem in Brazil, with a rate of modification by human activities of between 30 and 52 per cent, mainly due to damage caused by agriculture, livestock and road creation (Castelletti 2004; Leal et al. 2005). In addition, the region has been undergoing an intense process of environmental alteration and deterioration (Leal et al. 2003). Despite all these problems, it is one of the most neglected biomes, and is often placed in the background in discussions about the future of species and biodiversity conservation in the country (Velloso et al. 2001; MMA 2002). Only recently have governmental and non-governmental groups paid attention to this situation, because in addition to the need to minimise degradation, scientific knowledge is still insufficient (Velloso et al. 2001). This whole scenario, when combined with the history of low levels of economic and social development, has led to an intense process of environmental degradation, essentially due to poor management of natural resources (ICMBio 2011).

Most studies of the semi-arid region cover small areas, which inhibits the extrapolation of results, given the complexity of existing interactions (Leal et al. 2003; Tabarelli & Vicente 2004; ICMBio 2011; Albuquerque 2012). Consequently, there are major gaps in the data on the birds of the Caatinga, despite the fact that the avifauna is well studied in terms of taxonomy, natural history and geographical distribution (MMA 2002; Machado 2006; Roos et al. 2006).

The number of publications using species distribution modelling (SDM), not only applied to conservation studies, but also to ecological and evolutionary issues, is growing (Zim- mermann et al. 2010). There are still few studies using this tool in the Caatinga, and these are often fragmented, covering single species or small groups (e.g. Luiz 2010; Campos 2012; Oliveira & Cassemiro 2013; Mora- to et al 2014). Currently, the method has been successfully applied to answer biogeographic and conservation questions about endemic, rare or vulnerable birds in different biomes, such as the

Cerrado (e.g. Leite 2006; Marini et al. 2009; Marini et al. 2010; Ribeiro 2014) and the Atlantic Rainforest (e.g. Vieira 2007; Souza et al. 2011; Ferraz et al. 2012). However, these data are still insufficient and further research is needed to contribute to a complete understanding of the distribution patterns of key species.

Given the knowledge gaps in the Caatinga and the fact that it is a minimally protected natural region in Brazil, with around 7% of its territory in conservation units and less than 1% in integral protection units (MMA 1999; 2002; Leal et al. 2003; MMA 2014), this work aims to integrate known data for endemic birds, or those with little occurrence in adjacent biomes, creating potential distribution maps in a gradient of richness, in order to look for distribution patterns and zones of greater suitability for the species. The vegetation types in the areas of greatest richness will also be assessed, as well as the degree of representativeness of the Conservation Units (UCs) present, making this work relevant among other conservation studies for the biome.

The combination of lack of protection and the continuous loss of biological resources contributes to the extinction of species unique to the Caatinga (MMA 2007), where endemic taxa and threatened species deserve special attention due to the high vulnerability of the biota as a result of the current expansion of human activities (MMA 2002).

Actions in nature conservation compete with other areas, such as human occupation and extractivism (Neto 2013). Furthermore, the financial resources allocated to conservation are limited, so choosing the area to be preserved is a meticulous and strategic step (Neto 2013). From this study, it will also be possible to select priority areas for conservation, based on the regions of greatest richness. Previous studies have compared the most suitable areas for species with areas previously proposed for conservation (Leite 2006; Marini et al. 2009). In this respect, the regions proposed by The Nature Conservancy of Brazil & Associação Caatinga (2000), Biodi- versitas (2000), Velloso et al. (2001) and MMA (2002; 2004; 2007) will also be evaluated in order to corroborate them.

CHAPTER 2

Material and Methods

2.1. Study area

The Caatinga is the country's only exclusive biome, characterised by a wide geographical region with a high degree of vegetation variety, occupying around 10% of the country, with approximately 845,000 km^2 in northeastern Brazil (MMA 2014; IBGE 2014). It covers the states of Ceará, Rio Grande do Norte, Paraíba, Pernambuco, Piauí, Alagoas, Sergipe, Bahia, a small strip extending to the north of Minas Gerais, and another to the east of Maranhão (IBGE 2014). The name Caatinga comes from the Tupi-Guarani linguistic group, which means "white forest" (Prado 2003), alluding to the physiognomy of the vegetation which, during long dry periods, loses its leaves and displays bare, grey branches (Albuquerque & Bandeira 1995).

This region has a high evapotranspiration potential throughout the year (Sampaio 1995) and an extreme and irregular rainfall regime (Sick 1985). The biome has little annual variation in temperature, and the average annual and monthly temperatures are significantly high (Nimer 1972). The opposite occurs with rainfall, where annual precipitation is low, irregular and generally limited to a short period of the year, making it a determining factor for certain characteristics of the biome (Nimer 1972). In almost all of the caatinga, 50 to 70% of the annual rainfall is concentrated in three consecutive months, in which the rainfall regime is characterised by the existence of a relatively rainy season and a very dry period, where droughts last for at least six months (Nimer 1972). Generally, areas with six dry months correspond to a predominantly arboreal or transitional caatinga; those with seven to eight months, to predominantly shrubby vegetation and those with more than nine months, herbaceous, being more sparse in areas with 11 dry months (Nimer 1972). These phenomena of droughts and floods are frequent and have shaped the animal and plant life of the biome (Prado 2003).

Despite these harsh conditions, the biome is made up of a mosaic of phytophysiognomies that express climatic and relief variations, with the type of vegetation varying according to the soil and water availability, which are characterised as arboreal or shrub forests, comprising mainly trees and low shrubs, many of which have thorns, microphylia and some xerophytic characteristics, and are called "caatingas" (Velloso et al. 2001; Prado 2003; Leal et al. 2005; Alves 2007). Its typical landscapes correspond to lowlands, generally inter-plain depressions with shallow, stony soils (Zanella 2013), including tropical dry forests related to arboreal vegetation (Sampaio 1995).

The biome's vegetation is extremely diverse, including, in addition to the caatingas, associated environments called enclaves (Alves 2007). Different types of caatinga extend over higher regions and vary in relief, covering shrubby and arboreal caatinga, dry forest and moist forest, carrasco and open formations (Velloso et al. 2001). There are areas that, although floristically similar to the Caatinga vegetation, are not within its geographical limits, such as the dry valley of the Jequitinhonha river in Minas Gerais and certain regions of the Rio Grande basin in western Bahia (Prado 2003). On the other hand, the biome includes enclaves with a different physiognomy to the Caatinga, such as the Araripe plateau, with a characteristic Cerrado physiognomy, or other more humid areas, commonly referred to as

"high-altitude wetlands (Prado 2003). Due to this large mosaic, the
The Caatinga can be divided into eight ecoregions (Velloso et al. 2001).

Its fauna is shared with the Cerrado biome and adjacent humid forest areas (Sampaio 1995). The harsh climate and geomorphological characteristics of the region explain the existence of a flora with a considerable degree of endemism and xeric adaptations (Sampaio 1995).

The Caatinga complex stands out among the other physiogeographic provinces because it is apparently the richest in exceptional landscapes or enclaves, which makes its delimitation difficult (Fernandes & Bezerra 1990 apud Olmos et al. 2005). Many of its enclaves are due to geotopic factors, but others bear the marks of human activities, such as extensive cattle ranching and indiscriminate deforestation (Alves et al. 2009). Due to this polymorphism of the biome, one of the problems in defining endemic birds is determining the boundaries of the Caatinga itself (Olmos et al. 2005). It is possible that the variation in the relative abundance of bird species found in the Caatinga is more related to seasonality throughout the year (drought and rain) than to habitat complexity (Santos 2004).

2.2. Species selection

Firstly, species endemic to the Caatinga were selected, with less or no occurrence in adjacent biomes (Table 1). This list was compiled using Stotz (1996), Sick (1997), Pacheco (2004) and Olmos & Albano (2012) as references, totalling 32 species.

Table 1. List of species endemic to the Caatinga, with no or little occurrence in adjacent biomes and their respective references.

Species	Reference
Penelope jacucaca Spix, 1825	Pacheco (2004)
Anodorhynchus leari Bonaparte, 1856	Pacheco (2004), Olmos (2012)
Cyanopsitta spixii (Wagler, 1832)	Pacheco (2004), Olmos (2012)
Eupsittula cactorum (Kuhl, 1820)	Pacheco (2004), Olmos (2012)
Pyrrhura griseipectus Salvadori, 1900	
Hydropsalis vielliardi (Lencioni-Neto, 1994)	Pacheco (2004)
Hydropsalis hirundinacea (Spix, 1825)	Stotz (1996), Sick (1997), Olmos (2012)
Anopetia gounellei (Boucard, 1891)	Stotz (1996), Sick (1997), Pacheco (2004), Olmos (2012)
Augastes lumachella (Lesson, 1838)	
Picumnus pygmaeus (Lichtenstein, 1823)	Stotz (1996), Sick (1997), Pacheco (2004)
Picumnus fulvescens Stager, 1961	Stotz (1996), Sick (1997), Pacheco (2004)
Picumnus limae Snethlage, 1924	Stotz (1996), Sick (1997)
Myrmorchilus strigilatus (Wied, 1831)	
Formicivora iheringi Hellmayr, 1909	Stotz (1996)

Herpsilochmus sellowi Whitney & Pacheco, 2000	Olmos (2012)
Herpsilochmus pectoralis Sclater, 1857	Stotz (1996), Sick (1997)
Sakesphorus cristatus (Wied, 1831)	Sick (1997), Pacheco (2004), Olmos (2012)
Thamnophilus capistratus Lesson, 1840	Olmos (2012)
Hylopezus ochroleucus (Wied, 1831)	Pacheco (2004), Olmos (2012)
Xiphocolaptes falcirostris (Spix, 1824)	Pacheco (2004)
Megaxenops parnaguae Reiser, 1905	Sick (1997), Olmos (2012)
Pseudoseisura cristata (Spix, 1824)	Olmos (2012)
Synallaxis hellmayri Reiser, 1905	Stotz (1996), Sick (1997), Pacheco (2004), Olmos (2012)
Hemitriccus mirandae (Snethlage, 1925)	
Stigmatura budytoides (d'Orbigny & Lafresna- ye, 1837)	
Cyanocorax cyanopogon (Wied, 1821)	Sick (1997)
Icterus jamacaii (Gmelin, 1788)	Sick (1997)
Paroaria dominicana (Linnaeus, 1758)	Sick (1997), Pacheco (2004), Olmos (2012)
Sporophila albogularis (Spix, 1825)	Pacheco (2004), Olmos (2012)
Arremon franciscanus Raposo, 1997	Pacheco (2004)
Agelaioides fringillarius (Spix, 1824)	Olmos (2012)
Compsothraupis loricata (Lichtenstein, 1819)	Sick (1997)

Three species were excluded from the primary sources: Celeus obrie- ni, indicated by Stotz (1996), as it has a distribution restricted to the Cerrado biome (BirdLife International 2016); Herpsilochmus pileatus, also indicated by Stotz (1996), in which the taxonomic complex was recently revised, restricting this species to the coastal region of southern Bahia, referring to the Atlantic Forest domain (Whitney et al. 2000) and Rhopornis ardesiacus, proposed by Stotz (1996) and Sick (1997), because recently, Luiz, E. R. (2010) carried out a study on the geographical distribution and conservation of this species, in which he added more information about the vegetation formations related to its distribution, including occurrences referring to the Atlantic Rainforest domain, in addition, of the 12 localities surveyed by him, 11 belonged to the Atlantic Rainforest (Luiz 2010).

Five species were included in the list: Pyrrhura griseipectus, for being a species restricted to north-eastern Brazil in the state of Ceará (Forshaw & Cooper 1989; Collar 1997); Augastes lumachella for being a very common species in the plateau region between Morro do Chapéu, Andaraí and Barra da Estiva, being endemic to this region of Bahia (Sick 1997). Sibley & Monroe Jr (1990) and Schuchmann (1999) have already mentioned it also occurring in this plateau region in Minas Gerais; Myrmochilus strigilatus, as it is a typical species of the caatingas and corresponding forests, living in the closed and low forests of the Caatinga (Sick 1997). It occurs from north-eastern Brazil to Minas Gerais (Sibley & Monroe Jr 1990; Sick 1997; Zimmer & Isler 2003); Hemitriccus mirandae, as it is restricted to Ceará, Pernambuco and Alagoas (Sibley & Monroe Jr 1990) in forest enclaves that belong to the Caatinga domain (Sick 1997), and Stigmatura budytoides, which S. b. gracilis has a restricted occurrence in Pernambuco, northern Bahia and possibly southern Piauí, records for Minas Gerais have only recently been discovered (Fitzpatrick et al. 2004).

2.3. Occurrence data

The species had their occurrence data collected from the literature, visits to the scientific collection of the National Museum of Rio de Janeiro (MNRJ), the Museum of Zoology of São Paulo (MZUSP) and the SpeciesLink digital information system, developed by the team at the Centre for Reference and Environmental Information (CRIA), obtaining data from the Jacques Vielliard Neotropical Phonotheque (FNJV/UNICAMP), Professor Mello Leitão Biology Museum, Zoology Museum of the State University of Campinas, Science and Technology Museum of the Pontifical Catholic University of Rio Grande do Sul, Fundación Puerto Rastrojo of Colombia, Ecological and Economic Zoning of Acre and the Biota Environmental Information System (SinBiota/Fapesp) (splink.org.br). To complement the occurrence data, we used the WikiAves website, which is a digital platform designed to promote support for the online community of biologists and observers (WikiAves 2013). Previous studies have shown the reliability of this database and its importance as a complementary source of localities for bio-geographical studies (Santos 2012).

The geographical coordinates of the non-georeferenced localities were taken from the SpeciesLink network using the Geo-Loc tool. These coordinates were obtained from the IBGE source by selecting only the municipal centres, where they were all converted into decimal degrees, WGS84 datum.

2.4. Environmental variables

A principal component analysis (PCA) was carried out in order to select only the variables with low correlation. To do this, 19 digital files for the bioclimatic variables available on the *WorldClim* - Global Climate Data - website were run in *Rstudio* 3.1.1 using the Vegan, Dismo and Rgdal packages. In addition, an altitude variable was also extracted from WordClim. The climatic elements are restricted to the period 1950-2000 (Hijmans et *al.* 2005), and all the variables had a resolution of 5 arc minutes (c. 10km), due to the use of coordinates of municipal headquarters, in order to minimise the error made given the uncertainty of the animal's exact occurrence.

Variables with correlation values lower than 0.7 were used, and those with the highest value on the axis explaining most of the variation in the data were selected.

2.5. Algorithms and programmes

Four algorithms were used for modelling: BioClim, Mahalanobis Distance, Suported Vector Machines (SVM) and MaxEnt, the first three of which were run using the Open- Modeller 1.1.0 program and the last using the MaxEnt 3.3.3 software itself.

OpenModeller is a multiplatform tool with various algorithms included, developed by CRIA to carry out ecological niche modelling experiments (Munoz 2009). Some parameters for the Mahalanobis Distance were adjusted to create the models, such as the *"Metcic"* tool being changed to 2; "Maximum *dist.* " to 1000 and the "*Paarest* Tfpo/nf" to 0. The other algorithms - *BioClim* and SVM - were used as normal.

The *MaxEnt* algorithm was not run using the *OpenMo- deller* programme because the version was not up to date. Using the *MaxEnt 3.3.3 software,* the models were generated using the "Auto Fea- *tures"* settings and output in logistic format. The parameters used to generate the models were *"Random seed"* and "Remove *duplicate presence records".* The models were generated without *"Clampigg" and* with 1000 integers in order to improve their quality.

The *BioClim* algorithm predicts locations with favourable climatic conditions for the species by creating a climate profile from a set of environmental facts (Nix 1986 & Busby 1991 *apud* Hernandez *et al.* 2006). It is based on the calculation of a minimum enveloperetilinear in a multidimensional climatic space, and can be defined from a simple classification area (Guisan & Zimmermann 2000). The limits that define the envelope are obtained from the amplitude of each environmental variable in relation to the species' points of occurrence (Cpente *et al.* 1993). This algorithm uses only the points of occurrence, since the prediction is made without any reference to   or-tsmosts of the study area (Pearson 2007). The algorithm creates a distribution percentage for each variable, where in each *grid,* the values of each environmental variable are evaluated in order to determine its position in this percentage distribution (Hijmans & Graham 2006). This algorithm, therefore, does not produce a continuous gradient of habitat suitability, but it does indicate whether or not a site is suitable for the over-living of the species (Lima-Ribeiro & Diniz-Filho 2012).

An alternative to using rectilinear envelopes is the Mahalanobis distance. This is a multivariate method that classifies potential areas into a vector that expresses the average environmental conditions of all the records in the environmental space. This algorithm produces more accurate predictions of species distribution and can deal with correlations and interactions between climatic variables, i.e. it does not assume independence between variables (Farber & Kadmon 2003). These distance methods are the simplest representations of ecological niche logic, as they assume the existence of an ecological optimum for the survival of each species (Farber & Kadmon 2003; De Marco Jr & Siqueira 2009). This will be defined by the centroid of the environmental conditions related to the points of occurrence, establishing an ellipsoidal envelope in the environmental space, which will better reflect the species' response to the environmental gradient (Farber & Kadmon 2003; De Marco Jr & Siqueira 2009).

The SVM was introduced as a technique for solving pattern recognition problems; it minimises the risk of misclassifying patterns not yet seen by the probability distribution of the data (De Marco Jr & Siqueira 2009). It uses only the most informative data to generate potential distribution modelling, which makes it interesting in situations where the reliability of the species' occurrence records and/or environmental variables is incomplete (De Marco Jr & Siqueira 2009).

The MaxEnt algorithm focuses on how the environment where the species are known to occur relates to the entire environment of the study area ("background"). In this sense, the algorithm makes use of background data, i.e. it uses some of the occurrence points entered for the modelling as "background" data (Pe- arson 2007). The best approximation for an unknown probability distribution is the one that satisfies any restriction on the distribution, so it estimates the probability of a species being present by looking for the distribution that is closest to uniform (maximum entropy distribution) (Phillips et al. 2006). To do this, the algorithm endeavours to find a probability distribution that satisfies the specified restrictions. Based on the records of species occurrence, relating them to environmental conditions, it is possible to predict geographical areas with a higher degree of environmental suitability for the presence of species (Phil- lips et al. 2006).

2.6. Model evaluation

To allow for a more robust assessment of the models, the species that had a reasonable number of localities (over 29) had 70% of their occurrence points divided up as training points for model calibration, and 30% were used as test points for evaluation, as this calculates the model's hit percentage for each species. By inserting training and test points, the models generate confusion matrices to calculate the Area

Under the ROC Curve (AUC), which will be used to assess the models' performance. The matrix plots commission values, i.e. false presence, where the model shows suitability for the species, but there are no records of it; omission values, i.e. false absence, where there are records for the species in areas that the model did not indicate as potential for its occurrence; and true presence and absence values are also plotted (Fielding & Bell 1997; Pearce & Ferrier 2000). This matrix is used to calculate the sensitivity values, i.e. how accurate the model was with the occurrence data; and specificity, how accurate the model was with the absence data, where the sum of both values is the model's accuracy (Fielding & Bell 1997). The sensitivity values versus the complement of the specificity are inserted into the curve, generating the Area under the ROC curve, where the area under this curve is the so-called AUC, which will test whether the model classifies the presence with greater accuracy than a random prediction. The closer to 1 (one), the better the model, with 0.80 being considered good (Metz 1986). The ROC curve can also provide information for selecting an appropriate cut-off probability threshold (Pearce & Ferrier 2000).

2.7. Making the maps

As the study is aimed at conservation, pointing out possible areas for the preservation of species, a cut-off limit was chosen that minimised the chances of error, i.e. less inclusive models with low overprediction, which requires high cut-off values (Giannini et al. 2012). The cut-off used to generate the potential distributions was the Theresold ROC. Estimated by the models, it results in more conservative distribution maps, reducing the risk of selecting areas with lower suitability (Pearce & Ferrier 2000), generating a smaller optimum area, but with a higher degree of suitability.

The potential distribution maps for each species were generated by intersecting the different models in order to reduce the biases of each algorithm. These were overlaid to generate a potential map in a richness gradient, where regions with more than 19 species were considered to be areas of high richness, regions with 10 to 18 species as areas of medium richness and 0 to 9 species as low richness.

Due to the existence of several enclaves in the biome and the diverse phytophysiognomy, a vegetation shapefile was superimposed in order to correlate the areas of greatest richness with the vegetation types. This shape-file was obtained from the AMBDATA platform from the image maps produced by the RADAMBRASIL Project at a scale of 1:250,000, obtained from IBGE 1992 (AMBDATA).

The same was done with altitude maps, where a raster file extracted from the AMBDA- TA platform, generated from SRTM (Shuttle Radar Topographic Mis- sion) data, with a horizontal resolution of 3 arc-seconds (~90m) and vertical resolution (height) of 1 m (AMBDATA), was superimposed on the richness map. In addition, the potential distribution of richness was compared with other altitude maps (IBGE 2007; 2009; Oliveira & Medeiros 2012) in order to assess distribution patterns.

In addition, the wealth map was superimposed on the Conservation Units System in order to investigate their representativeness and indicate possible new areas for the implementation of protection areas. This database of Conservation Units corresponds to a compilation made by the Environmental Zoning Coordination of the IBA-MA (www.ibama.gov.br/zoneamento-ambiental) based on data from official and unofficial sources.

2.8. Priority areas for conservation

Taking into account the preservation of the Caatinga's endemic birds, the areas and municipalities that obtained a high degree of potential richness, i.e. above 19 species, were considered to be important regions that deserve to be highlighted for the conservation of the biome.

In addition, previous proposals for priority areas were analysed (The Nature Conservancy of Brazil & Caatinga Association 2000; Biodiversitas 2000; Velloso et al. 2001; MMA 2002; 2004; 2007), comparing these proposals to the regions of greatest wealth, in order to corroborate their importance.

CHAPTER 3

Results

A total of 4,644 occurrence records were obtained, covering 1,406 geographical locations. The most significant source of data was the literature with 961 records, followed by the MZUSP bird collection with 244, the SpeciesLink network with 112 and the MNRJ Ornithology Sector collection with 97. Data from the WikiAves platform complemented the study with 3152 records. The number of geographical coordinates obtained from training and testing for the modelling of each species are shown in the table below (Table 2).

Table 2. List of species selected for this study with their respective number of sites used for model calibration (No. of sites/Training) and testing (No. of sites/Testing).

Species	N° locations/ Training	N° locations/ Test
Penelope jacucaca	28	11
Anodorhynchus leari	12	-
Cyanopsitta spixii	8	-
Eupsittula cactorum	206	88
Pyrrhura griseipectus	4	-
Hydropsalis vielliardi	6	-
Hydropsalis hirundinacea	66	28
Anopetia gounellei	49	21
Augastes lumachella	21	8
Picumnus pygmaeus	103	44
Picumnus fulvescens	25	10
Picumnus limae	45	19
Myrmorchilus strigilatus	115	48
Formicivora iheríngi	16	-
Herpsilochmus sellowi	75	31
Herpsilochmus pectoralis	25	10
Sakesphorus cristatus	86	36
Thamnophilus capistratus	126	54
Hylopezus ochroleucus	35	14
Xiphocolaptes	21	9

falcirostris		
Megaxenops	52	21
parnaguae		
Pseudoseisura	196	83
cristata		
Synallaxis hellmayri	42	18
Hemitriccus mirandae	20	-
Stigmatura budytoides	48	20
Cyanocorax		
cyanopogon	268	114
Icterus jamacaii	378	162
Paroaria dominicana	404	173
Sporophila albogularis	215	91
Arremon franciscanus	17	-
Agelaioides fringillarius	166	70
Compsothraupis loricata	140	59

Of all the species, three were excluded from the analysis because they did not obtain the minimum number needed for modelling, with fewer occurrence points than environmental variables. These were Cyanopsitta spixii with eight records, Hydropsalis vielliardi with six and Pyrrhura griseipectus with four.

3.1. Selection of environmental variables

As a result of the PCA, of the 19 environmental layers inserted, 11 were highly correlated (Appendix A), where only eight variables were selected for species modelling: Mean diurnal variation (Bio 2), Isothermality (Bio 3), Seasonal temperature (Bio 4), Mean temperature of the coldest quarter (Bio 11), Annual precipitation (Bio 12), Seasonal precipitation (Bio 15), Precipitation of the driest quarter (Bio 17) and Precipitation of the coldest quarter (Bio 19). The altitude variable (Bio 0) was also added to the modelling due to its importance for species endemic to the Caatinga, as these have been associated with higher altitude regions in previous tests.

3.2. Wealth models and map

The potential distribution maps generated by the Bio- Clim, SVM and MaxEnt algorithms had significant AUCs above 0.92. However, for the majority of species, the maps generated by Mahalanobis Distance had a low evaluation value, including Anodorhyn chus leari, Formicivora ihering and Arremon fransiscanus with an AUC of 0.50; Penelope jacucaca, Anopetia gounellei, Myrmorchilus strigilatus, Herpsilochmus sellowi, Sakesphorus cristatus, Hylopezus ochroleucus, Megaxenops parnaguae, Synallaxis hellmayri, Paroaria dominicana and Compsothraupis loricata with an AUC of approximately 0.60; and Eupsittula cactorum, Hydropsalis hirundinacea, Picumnus pygmaeus, Picumnus fulvescens, Picumnus limae, Herpsilochmus pectoralis, Thamnophilus capistratus, Pseudoseisura cristata, Hemitriccus miran- dae, Stigmatura budytoides, Sporophila albogularis, Agelaioides fringilla- rius, Cyanocorax cyanopogon and Icterus jamacaii with around 0.70 (Appendix B). An intersection of the different models was made using only the algorithms that performed well, i.e. with an evaluation value above 0.80 (Appendix C).

The means and variances of the AUCs, respectively, were as follows: BioClim 0.9842069 and 0.0002417; Mahalanobis Distance 0. 6945517 and 0.006624756; SVM 0.989069 and 2.863793e-05; and MaxEnt 0.99 15862 and 5.010837e-05 (Graph 1 and 2).

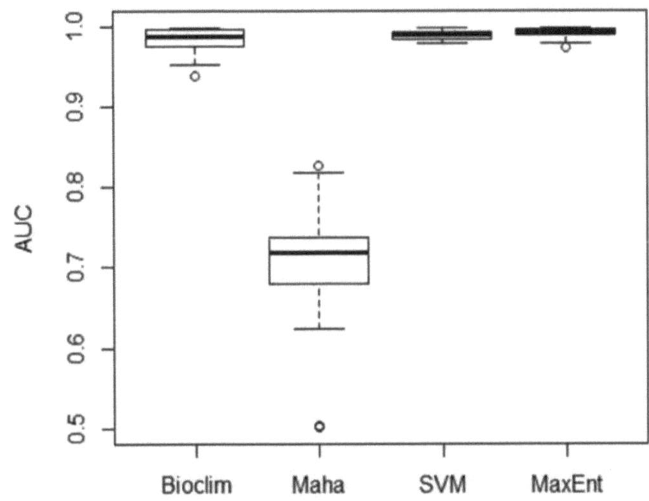

Graph 1. Boxplot for the AUCs of the four models, showing that the Mahalanobis Distance presented the greatest variance between the results and the lowest performance.

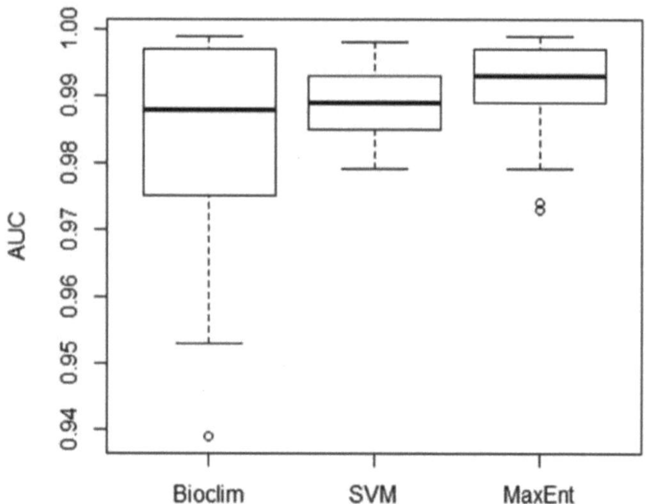

Graph 2. Boxplot for the AUCs of the three best models, BioClim, SVM and MaxEnt, showing that the latter performed best.

The potential richness map (Figure 1) showed areas suitable for a maximum of 23 species, pointing to two large regions of greater richness, one spot to the north and another to the south. When compared with altitude maps (IBGE 2007; 2009; Oliveira & Medeiros 2012), it was observed that these regions are associated with higher altitude areas up to around 850m, with richness tending to decrease above this threshold. In the northern region, richness increases from 400m on average, while in the southern region it tends to increase at around 500m. The former is associated with the Chapada do Araripe and the Borborema Plateau, while the latter corresponds to the Upper São Francisco region, between the slopes of the Serra Geral de Goiás and the Serra do Espinhaço, as well as a strip around the Chapada da Diamantina.

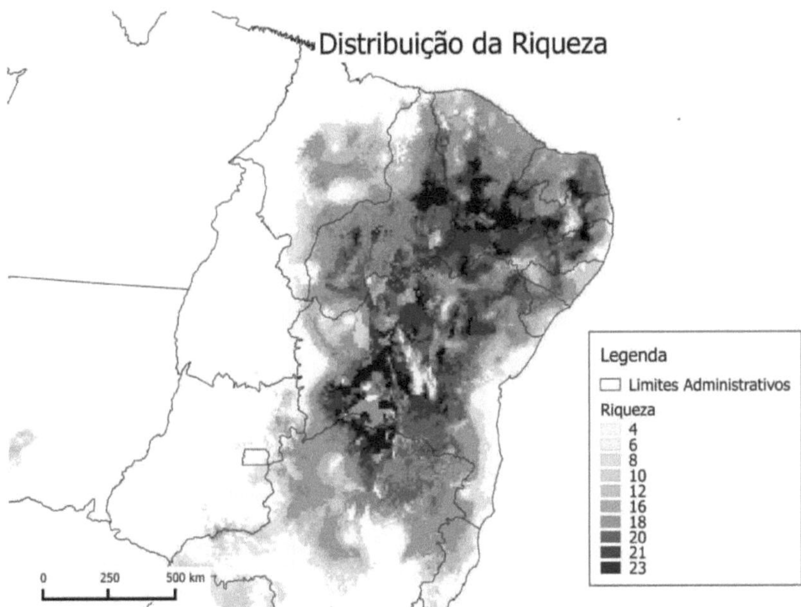

Figure 1. Distribution of potential richness, showing two large zones of greater richness, one to the north, referring to the Araripe Plateau and the Borborema Plateau, and the other to the south, corresponding to the Upper São Francisco region.

Analysing the distribution of the vegetation, it was observed that the areas of greatest richness refer to Seasonal Deciduous and Semi-Deciduous Forest, with deciduous secondary vegetation and agricultural activities, Steppic vegetation, areas of ecological tension between Steppic vegetation and Seasonal Forest, as well as areas of open tree Savannah.

It was found that 309 municipalities in the northeast have a high degree of potential richness, i.e. suitability for at least 19 of the 29 endemic species modelled (Appendix D). This shows that they deserve attention when it comes to conserving the endemic avifauna of the caatinga.

3.3 Conservation units

By overlaying the shapefile of the Conservation Units, it was observed that there are very few that representatively cover the areas of greatest richness. Of the 32 Integral

Protection Units in the bio- ma, 26 are exclusive to the Caatinga and of these only five cover small areas of greater richness with any representativeness. There are also two units in the ecotone between the Caatinga and the Cerrado and two that are already within the Cerrado biome (Table 3).

Table 3. List of Full Protection Conservation Units and their respective municipalities, states and biomes that cover areas with high potential richness of modelled avifauna.

Conservation Unit	Municipalities	State	Biome
Raso da Catarina Ecological Station	Jeremoabo, Rodelas, Paulo Afonso	BA	Caatinga
Jaíba Biological Reserve	Matias Cardoso	MG	Caatinga
Serra da Capivara National Park	João Costa, Coronel José Dias, São Raimundo Nonato	PI	Caatinga
São Francisco River Natural Monument	Canindé de São Francisco, Delmiro Gouveia, Olho D'Água do Casado, Piranhas and Paulo Afonso	BA, SE, AL	Caatinga
Mata Seca State Park	Itacarambi	MG	Caatinga
Lagoa do Cajueiro State Park	Matias Cardoso	MG	Caatinga/ Cerrado
Peruaçu Caves National Park	Januária, Itacarambi, São João das Missões	MG	Caatinga/ Cerrado
Wildlife Refuge Veredas do Oeste Baiano	Jaborandi, Coconuts	BA	Cerrado
Serra Azul Biological Reserve	Mateus Leme, Igarapé, Itaúna	MG	Cerrado

It was also noted that of the 38 Sustainable Use Conservation Units in the Caatinga, 22 are exclusive to the biome. Where seven exclusive units and one in a transition area between the Cerrado and Caatinga cover regions of high richness (Table

4).

List of Sustainable Use Conservation Units and their respective municipalities, states and biomes that cover areas with high potential richness of modelled avifauna.

Conservation Unit	Municipalities	State	Biome
Gruta dos Brejões/Veredas do Romão Environmental Protection Area Gramado	Morro do Chapéu	BA	Caatinga
Serra Branca/Raso da Catarina Environmental Protection Area	Jeromoabo	BA	Caatinga
Marimbu/Iraquara Environmental Protection Area	Abaíra, Piatã, Rio de Contas	BA	Caatinga
Contendas do Sincorá National Forest	Contendas do Sincorá, Tanhaçu	BA	Caatinga
Serra do Orobó Area of Relevant Ecological Interest	Rui Barbosa, Itaberaba	BA	Caatinga
Negreiro National Forest	Serrita and Parnamirim	PE	Caatinga
Chapada do Araripe Environmental Protection Area	-	CE, PE, PI	Caatinga
Lajedão Environmental Protection Area	Matias Cardoso	MG	Caatinga/Cerrado

The Chapada do Araripe Environmental Protection Area is part of an enclave of medium richness, with suitability for the presence of 15 to 18 species and small regions with suitability for 19 to 23 species. The surrounding region is more representative of areas with a high degree of richness.

CHAPTER 4

Discussion

According to IBGE data, the urbanisation rate in Brazil's northeastern region reached 60% in the 1990s and approximately 74% in 2010. This rapid process brings with it a series of activities that, when not structured correctly, cause irreparable damage to the environment, including the loss of biodiversity (Fernandes & Medeiros 2009). Given this critical panorama, it is essential that the results of different studies on the Caatinga biome and its interfaces are available as soon as possible. Currently, it is difficult to obtain information about biodiversity. There is a considerable amount of data, but it is often dispersed and with restricted access (Canhos 2003). In this sense, there is a clear need for efforts to build a network of useful and easily accessible information, which can be achieved through advances in communication and technology (Canhos et al. 2001). In the case of this work, it was difficult to find and compile the geographical coordinates, which reinforces the importance of making digital databases available, such as the SpeciesLink network.

4.1. Potential distribution models and maps

Species distribution models have been used with relative success to investigate various issues in the field of conservation, ecology, biogeography and phylogeography, although each method has its limitations (Guisan & Thuiller 2005). The models generated by the Mahalanobis distance for species with few records, such as Anodorhychus leari, Formicivora ihering and Arre- mon fransiscanus, with 12, 16 and 17 points respectively, had extremely low AUCs (0.50). This algorithm takes into account a covariance matrix between the environmental variables at the points of occurrence (Farber & Kadmon 2003), which allows the model to be interpreted as an expression of the environmental restrictions suffered by the species, making it unsuccessful for modelling rare species (De Marco Jr & Siqueira 2009). As the evaluation value was minimal for these three species, the consensus maps from BioClim, SVM and MaxEnt were used.

It can be seen that for species with less than 20 points of occurrence, the Mahalanobis model evaluation values were extremely low (0.50), i.e. these models had the same significance as a random prediction. For taxa with more than this threshold of records entered for modelling, the Mahalanobis distance showed slightly more satisfactory results, but in some cases, even with a reasonable number of georeferenced points, the AUC was low, ranging from 0.62 to 0.76, which also made it impossible to use it for these species to generate the final potential distribution map.

Another restriction of the models is the factors that determine the distribution of the species. There is a range of these factors, not all of which are analysed by the algorithms (Soberón & Peterson 2005). It is assumed that for the species to be present at a given point, three conditions must be favourable: dispersal capacity (M), either by the individual's own movements or by the dispersal of the species by external agents, determining the regions accessible to it; the abiotic environment (A), i.e. environmental conditions favourable to the creation, survival and reproduction of individuals; and the biotic environment (B) made up of species interactions, such as competition, predation, pathologies, etc.., associated with the dynamics and availability of resources (Gui- san & Thuiller 2005; Soberón & Peterson 2005). In order to estimate the actual distribution, hypotheses are required about the degrees of overlap between A, B and M, but modelling algorithms generally only estimate A (Soberón & Peterson 2005).

These algorithms tend to extrapolate the data in order to identify areas of presence based on associations between specific occurrences and environmental data (Soberón & Peterson 2005). According to the environmental variables entered, the regions are determined by the algorithm based on ecological similarities with the areas where the species is known to occur, so the modelling algorithms only find regions that are similar in terms of layers to where the points of occurrence are located (Soberón & Peterson 2005).

It is possible that the models found similarities in the environmental variables inserted for some regions with the species' areas of occurrence, being calculated in different ways and causing the potential distribution maps to be different from those with the known current distribution, presenting new areas with suitability for its occurrence.

Sakesphorus cristatus, Cyanopogon cyanocorax, Icterus jamacaii and Paroaria dominicana were examples of species where the potential distribution generated by the consensus of the four algorithms showed wider distributions than those currently known.

Sakesphorus cristatus is known to be distributed in the centre of Ceará, the south-east of Piauí, the west of Pernambuco, Bahia, except along the coast, and the north of Minas Gerais (Zimmer & Isler 2003), a region where its distribution has been extended (Marini & Lopes 2005). However, the potential distribution included a portion to the south-east of Maranhão, as well as eastern Goiás and a strip to the north of Mato Grosso do Sul and southern Mato Grosso, where there are no records for the species. It is possible that the species does not occur in these environments in which the model showed suitability due to the presence of geographical barriers, specific habitat requirements, or other factors such as interactions with other species. S. cristatus is known to be common in medium-sized vegetation environments, with an interlaced canopy and continuous forest cover of around 4 to 5 metres in height, such as in liana forests (Zimmer & Isler 2003).

Cyanocorax cyanopogon is distributed from Maranhão eastwards to Ceará and Paraíba, down to the south-east of Pará, the far east of Mato Grosso, Goiás, Minas Gerais, Bahia and an isolated population in Espírito Santo (dos Anjos 2009), which arrived in the region due to deforestation (Sick 1997). The model may have found climatic similarities between the regions with occurrence records for the species in Rondônia, Mato Grosso, except for the extreme north, northern Mato Grosso do Sul and northern São Paulo, presenting potential for the occurrence of C. cyanopogon. All states have records for the species, except for Mato Grosso do Sul. C. cyanopogon is known to inhabit areas of mainly dry vegetation ranging from 400m to 1100m and is commonly seen in riparian forest, secondary forest and the edge of tropical deciduous forests (dos Anjos 2009). The suitability indicated by the modelling for Mato Grosso do Sul may have been due to the models not taking into account biotic or dispersal factors. The absence of the species in these regions is due to the presence of its sister species, C. chrysops, where they are geographically excluded (Sick 1997). Although they are related species, they show a parapatric distribution and can be considered allospecies, i.e. allopatric species with similar morphological and ecological characteristics, which together form a superspecies (Sick 1997). In Brazil, C. chrysops is distributed from Mato Grosso do Sul and southern São Paulo to Rio Grande do Sul (dos Anjos 2009).

Icterus jamacaii is distributed from Maranhão and Tocantins eastwards to Rio Grande do Norte and Paraíba, southwards to northern Minas Gerais (Fraga 2011). Its potential distribution includes the entire state of Minas Gerais, northern Rio de Janeiro, northern São Paulo, a strip of eastern Goiás, northern Mato Grosso do Sul and southern Mato Grosso. These regions already have records for the species. It is a species that lives in clearings and edges of primary or secondary caatinga, in dry forests (Fraga 2011) and is expanding its area of occurrence.

Paroaria dominicana has its current known distribution in north-eastern Brazil, from southern Maranhão, Piauí and Ceará to eastern Goiás and northern Minas Gerais (Jaramillo 2011). It is expanding its distribution into new areas such as southern Bahia and Minas Gerais (Sick 1997). Couples that have escaped from captivity can breed in the wild, occupying other areas (Sick 1997), which has led to their distribution expanding. The potential distribution includes the centre-south of Maranhão to Rio Grande do Norte, as well as the state of São Paulo, northern Mato Grosso do Sul and southern Mato Grosso. The latter two do not yet have records for the species. These are potential areas for the occupation of P. dominicana, which shows that care must be taken to avoid the reintroduction of this species in these locations.

The other species had potential distributions similar to the limits indicated by their known current distribution.

Penelope jacucaca: A vulnerable species (Silveira & Straube 2008; IUCN 2013), it is distributed in Maranhão, Piauí, Ceará, Paraíba, Alagoas and Bahia (Sick 1997). Its potential distribution map showed suitability for Piauí, a portion of Ceará, Rio Grande do Norte, Paraíba, Pernambuco, Alagoas, Sergipe, down through the central region of Bahia to northern Minas Gerais, where all these regions have records for the species.

Anodorhynchus leari: With a restricted distribution and critically endangered (Silveira & Straube 2008; IUCN 2013), it only occurs in the caatinga with thorny shrub vegetation, mainly in areas where the Licuri palm (Syagrus coronata), its main food, occurs. It also inhabits areas close to canyon walls which it uses for breeding and living (Collar 1997, Sick 1997). It is only known in the north-east of Bahia (Forshaw & Cooper 1989), in the municipalities of Canudos, Euclides da Cunha, Jeremoabo, Paulo Afonso, Uauá, Campo Formoso and Sento Sé and Santa Brígida (Santos Neto & Camandaroba 2007). Its potential distribution has also proved to be restricted, occupying just one patch in the north of Bahia.

Eupsittula cactorum: Distributed in north-eastern Brazil (Forshaw & Cooper 1989, Collar 1997, Sick 1997). Its potential distribution is from Piauí, eastwards to Rio Grande do Norte and southwards to the centre of Minas Gerais. It occupies the coast except for the state of Bahia and has a distribution patch in the centre of Maranhão, where there are no records for the species.

Hydropsalis hirundinacea: It is distributed in north-eastern Brazil, from the north of Ceará to the extreme north of Bahia, from the south of Piauí to Bahia and Alagoas, and in eastern Brazil in Espirito Santo (Cleere 1999). Its potential distribution is from Piauí eastwards to Rio Grande do Norte, occupying central Bahia, northern Minas Gerais and northern Espírito Santo, where all these regions already have records for the species.

Anopetia gounellei: It ranges from Piauí to Bahia (Schuchmannm 1999), including Ceará (Sick 1997). Its potential distribution covered the north-east of Brazil, except for the coast. It occupies the east and south of Piauí, Ceará, Rio Grande do

Norte, Paraíba, Pernambuco, a strip in central Alagoas, western Sergipe and central Bahia up to northern Minas Gerais. All states have records of occurrence for the species.

Augastes lumachella: This is a near threatened species (IUCN 2013) restricted to the mountain top areas of the Espinhaço Chain (Vasconcelos et al. 2008), where its current distribution covers the northeast and centre of Bahia and Minas Gerais (Schuchmann 1999), recently Souza et al. (2009) expanded the occurrence of the species to the northern portion of the Serra do Espinhaço, in the north of Bahia. Its potential distribution showed suitability for the centre of Bahia.

Picumnus pygmaeus: It is distributed in the centre of Maranhão and Piauí, eastwards to Pernambuco and southwards to the north-east of Goiás and the extreme north of Minas Gerais (Winker & Christie 2002). It has shown areas of suitability within these limits, as well as Rio Grande de Norte, Paraíba and Espírito Santo.

Picumnus fulvescens: Near Threatened (IUCN 2013) has a known distribution in eastern Piauí along southern Ceará to Paraíba, Pernambuco and Alagoas (Winkler & Christie 2002). Its potential distribution only included eastern Rio Grande do Norte to northern Sergipe, where it already has records for the species.

Picumnus limae: An endangered species according to Silveira & Straube (2008), with a current distribution in the interior of Ceará to the west of Paraíba (Short 1982). Its potential distribution differed only in the far east of Piauí, showing suitability for the region.

Myrmorchilus strigilatus: It is distributed in north-eastern Brazil, to the extreme east of Piauí, the centre of Ceará, and the south of Rio Grande do Norte to the north of Minas Gerais (Zimmer and Isler 2003). Its potential distribution showed suitability in these regions, to the east and south of Piauí, Ceará, with the exception of the north, Rio Grande do Norte, Paraíba, Pernambuco, down through the centre of Bahia to the centre of Minas Gerais.

Formicivora iheringi: According to the IUCN (2013) it is close to endangered, with a distribution in eastern Brazil, in the eastern interior of Bahia and north-eastern Minas Gerais (Zimmer & Isler 2003). Its potential distribution showed areas of suitability in the north of Minas Gerais with patches of distribution in the centre and south of Bahia, as well as small sites in the north of Pernambuco, where there is no record of the species, probably due to the presence of geographical barriers. It lives in Chapadas and mountain slopes no lower than 600m (Sick 1997).

Herpsilochmus sellowi: It is distributed in the centre-north and east of Brazil, from Maranhão to Rio Grande do Norte and as far north as Minas Gerais (Zimmer & Isler 2003). Its potential distribution extends from the south-east of Ceará to the east of Rio Grande do Norte and Paraíba, continuing south to the centre of Minas Gerais.

Herpsilochmus pectoralis: Vulnerable species (Silveira & Straube 2008; IUCN 2013), with a distribution from northeastern Maranhão to eastern Rio Grande do Norte, in the direction of Sergipe, to northeastern Bahia (Zimmer & Isler 2003). Its potential distribution covered the same regions, including a spot in the north of Espírito Santo. The models probably found climatic similarities between the regions where H. pectoralis occurs and Espírito Santo, but there are no records for the species in these areas.

Thamnophilus capistratus: According to the literature, it is found in the east and south of Piauí, Ceará and Rio Grande do Norte, heading south to the extreme north of Minas Gerais and the centre of Bahia (Zimmer & Isler 2003). The species showed suitability for the centre of Maranhão, the entire state of Piauí, Ceará, Rio Grande do Norte, Paraíba, Pernambuco, Alagoas, Sergipe, Bahia, northern Minas Gerais and western Espirito Santo. There are records for the species in all localities except Espírito Santo.

Hylopezus ochroleucus: Near threatened (IUCN 2013), it is distributed in the west and south of Ceará, Piauí, eastern Pernambuco, Bahia and the extreme north of Minas Gerais (Krabbe & Schulenberg 2003). Its potential distribution covered the same regions already known to occur for the species.

Xiphocolaptes falcirostris: This is a vulnerable species (Silveira & Straube 2008; IUCN 2013) with two subspecies that replace each other (Sick 1997). Xipocolaptes falcirostris falcirostris, with a distribution in the east of Maranhão and Ceará to the west of Paraíba and Pernambuco in a southerly direction to the north-east of Bahia. While X. f. franciscanus is distributed west of the São Francisco River, in western Bahia and north and north-western Minas Gerais (Marantz et al. 2003). The potential distribution of X. falcirostris showed suitability for regions including the far west of Rio Grande do Norte and the far north of Goiás, where there was one record of the species. There were no records of occurrence in Rio Grande do Norte.

Megaxenops parnaguae: Its known distribution is in eastern Brazil, from southern Piauí, Ceará and western Pernambuco to western Bahia and north-western Minas Gerais (Remsen 2003). The potential distribution covered the same localities, including only the far west and south of Paraíba, where no records were found.

Pseudoseisura cristata: Its known distribution covers eastern Maranhão, Paraíba and Pernambuco, down to the centre of Minas Gerais (Remsen 2003). Its potential distribution covers eastern Piauí, Ceará, Rio Grande do Norte, Paraíba, Pernambuco, Alagoas, Sergipe, Bahia, northern Minas Gerais and Espírito Santos, where only the latter has no occurrence records for the species.

Synallaxis hellmayri: According to IUCN (2013), it is close to endangered, being distributed in north-eastern Piauí, western Pernambuco, northern Bahia, and an isolated patch in the far north of Minas Gerais (Remsen 2003). Its potential distribution covered the same areas, including western Paraíba and a strip extending to the centre of Bahia and the far north of Minas Gerais. All localities have records for the species.

Hemitriccus mirandae: An endangered species according to Silveira & Straube (2008) and vulnerable according to IUCN (2013). It is distributed in north-eastern Brazil, on isolated ridges in northern Ceará, Paraíba, Pernambuco, Alagoas and southern Bahia (Fitzpatrick et al. 2004). Its potential distribution corresponded to its current known distribution, showing suitability in isolated patches of distribution, including a patch in Sergipe, Rio Grande do Norte and another in northern Piauí. No records were obtained for the species in Sergipe and Rio Grande do Norte. It is likely that H mirandae is not found in these locations due to the presence of geographical barriers.

Stigmatura budytoides: It is currently distributed in Pernambuco, northern Bahia and possibly southern Piauí, with a record in Minas Gerais (Fitzpatrick et al. 2004). Its potential distribution includes the south-east of Piauí, south of Ceará and Paraíba, with a strip going up to the east of Rio Grande do Norte, occupying the state of Pernambuco, except for the coast, west of Alagoas and Sergipe, down through the centre of Bahia and north of Minas Gerais, as well as northern Espírito Santo. All regions except Espírito Santo have occurrences of the species.

Sporophila albogularis: It is distributed in the north-east of Brazil, from Piauí eastwards to Pernambuco and down to the north of Bahia (Jara- millo 2011). Its potential distribution includes the centre of Maranhão and Espirito Santo, where there has been one record.

Arremon franciscanus: Near Threatened (IUCN 2013) has a restricted distribution in southern Bahia and northern Minas Gerais (Jaramillo 2011), corresponding to the map generated by the potential distribution.

Agelaioides fringillarius: It is currently distributed in Brazil's northeastern interior, occupying Piauí, Ceará, Rio Grande do Norte, extending southwards to northern Minas Gerais (Fraga 2011). Its distribution generated by the models included the central portion of Maranhão and Espirito Santo, where there are no records for the species.

Compsothraupis loricata: Its current known distribution encompasses eastern Maranhão, Piauí, Ceará, Pernambuco and Alagoas, continuing south to central-western Goiás and northern Minas Gerais (Hilty 2011). Its potential distribution includes the west of Espírito Santo, where there are no records for the species.

4.2. Wealth map

There are few studies that present richness distribution maps for the entire Caatinga, demonstrating that greater efforts should be made to understand the distribution patterns of the taxa in this biome on a broad scale. Previous studies have identified some areas of richness that overlap with this study. Such as the diversity of bees, which pointed to the Wooded Steppe Savannah with the highest richness (Moura 2003) and the diversity of snakes, also pointing to areas of contact with neighbouring morphoclimatic domains and the semi-arid caatinga (Guedes et al. 2014).

When analysing the areas of greatest richness in this study, it can be seen that these are equivalent to higher altitude areas from around 450m with a maximum limit of 850 metres on average. Due to the strong solar radiation that the Caatinga region suffers, the lower areas have the highest annual temperature averages, with only the areas above 200 to 250 metres having a milder annual average temperature of less than 26°C. In addition, due to the influence of the trade winds, the eastern coast is less hot, ranging from 26°C to 24°C. In areas where the levels are above 850-900m in at least one month of the year, the average temperature is below 18°C (Nimer 1972). In this context, it can be seen that richness tends to be higher at milder temperatures, but not as low. The influence of the trade winds, combined with the altitude, makes the Diamantina surface, above 700-950m, and the Borborema, above 600-850m, the areas with the mildest temperatures in the region (Nimer 1972), where richness tends to decline. Although the Borborema Plateau showed the greatest richness, this region varies between 150-650m in altitude with peaks reaching up to 1000m (Velloso et al. 2001), where above around 850m the richness tends to decrease.

Regions above 1000m in Diamantina and 950m in Bor- borema record annual averages of less than 20°C (Nimer 1972). These characteristics corroborate that the endemic species of the caatinga have a temperature optimum, showing that richness increases in a certain temperature range and tends to decrease as altitude increases and, consequently, temperature decreases. This average annual temperature range has maximum values below 26°C *and* minimum values above 20°C, with the average being around 23°C.

It was also realised that the greatest richness corresponds to areas of Seasonal Deciduous and Semi-Deciduous Forest, the arid northeastern hinterland with its steppe vegetation, the contact areas between these two domains, as well as areas of open tree Savannah.

4.3. Conservation units and priority areas

It was observed that these areas with the greatest richness of endemic birds, i.e. suitable for at least 19 species, lack protected areas, making it necessary to implement Conservation Units in these regions. In addition, many of the existing Conservation Units have various problems, including the lack of a management plan and a lack of staff prepared to maintain them, suggesting the need to make the Units that have already been decreed functional (The Nature Conservancy of Brazil & Associação Caatinga 2000).

In order to conserve this diversity, the area with the greatest richness, making it a priority for conservation, corresponds to the region around the Chapada do Araripe, with 19 to 23 species, followed by a small region to the east of Paraíba on the border with Pernambuco, which is suitable for 20 to 23 species, some small portions to the south of Bahia, with 19 to 22, a region to the far north of Minas Gerais, with suitability for 20 to 21 species and finally, another small area to the east of Rio Grande do Norte on the border with Paraíba, with suitability for 19 to 21 taxa (Figure 2).

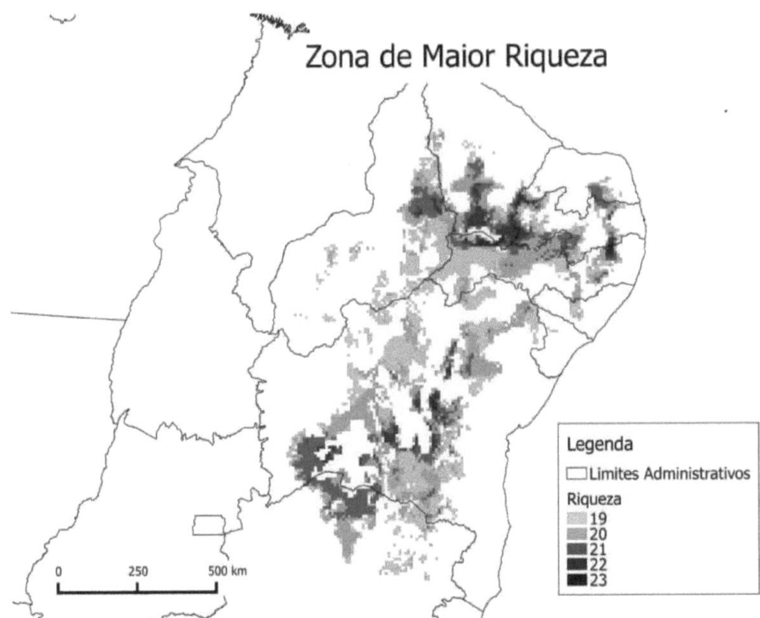

Figure 2: Zones of greatest richness with suitability ranging from 19 to 23 species. These deserve to be highlighted for the development of conservation actions.

Analysing previous proposals (The Nature Conservancy of Brazil & Associação Caatinga 2000; Biodiversitas 2000; Velloso et al. 2001; MMA 2002; 2004; 2007), it can be seen that 83 areas out of a total of 198 (42%) cover regions with a high degree of richness, i.e. over 19 species, corroborating the importance for conservation of the area in terms of preserving endemic birds. All the corroborated areas have been listed, indicating the number of species the region potentially harbours and their percentage in the territory (Appendix E).

Of the eight priority areas proposed by The Nature Conservancy of Brazil & Associação Caatinga (2000), this study corroborated the relevance of two, referring to the areas of greatest richness. Biodiversitas (2000) proposed 57 regions, of which 29 were corroborated; while Velloso et al. (2001) selected 17 areas from those proposed by Biodiversitas with the greatest importance for conservation, of which six were corroborated. MMA (2002) proposed 35 areas, 16 of which correspond to areas of greatest richness, MMA (2004) proposed 28 areas for the creation of Conservation

Units, 16 of which were corroborated, and MMA (2007) proposed 94 areas, 35 of which were corroborated.

Some of these areas deserve special attention for conservation, such as the Blue Macaw breeding area proposed by MMA (2004), located in Curaça, a municipality also proposed by Biodiver- sitas 2000, and Juazeiro, BA. It harbours a potential 20 to 22 species in half of its territory, also demonstrating its importance for the conservation of endemic avifauna. The type of vegetation found in this area is essential for the survival of Cyanopsitta spixii, a species that is possibly extinct in the wild. Since 2000, no free-living specimens have been recorded (BirdLife In- ternational 2013), and its threat status is considered critically endangered according to IUCN 2013. In addition, there is little representation of this phytophysiognomy in existing UCs (MMA 2004), demonstrating once again the great importance of conserving this area.

Raso da Catarina, BA, is also worth mentioning. This area, proposed by Biodiversitas (2000) and MMA (2002), has less than 50 per cent of its territory suitable for 19 to 20 species, but although it has vast regions with low and medium levels of richness, potentially harbouring nine to 18 species, it is the area of occurrence of Ano- dorhynchus leari (Amaral et al. 2005; Silveira & Straube 2008), a globally endangered species (IUCN 2013) and critically endangered according to Silveira & Straube (2008).

Some other regions that don't show high richness, even in part of the territory, also deserve to be highlighted because they are potential areas for the occurrence of Hemitriccus mi- randae and Picmnus limae, endangered species (Silveira & Straube 2008) (appendix F).

Of the municipalities that showed a high degree of potential richness, 183 were not included in the priority areas proposed for conservation consulted, but they deserve to be highlighted and strategies suggested for the preservation of this diversity of endemic avifauna (Appendix G).

4.4. Future prospects

Future work should include a greater number of species, especially the threatened and rare ones, as well as covering other taxa, enabling a greater understanding of species distribution patterns. It is also interesting to look at the similarity between the existing UCs and the areas to be conserved, with the aim of creating protection strategies in environments that harbour poorly protected species, in order to ensure the entire biodiversity of the Caatinga, especially the exclusive and threatened taxa.

In general terms, it will only be possible to safeguard the biodiversity of the Caatinga through efforts supported by different studies that prove the uniqueness of evolutionary, biogeographical and natural history patterns.

CHAPTER 5

Conclusions

- The Caatinga has two regions of greater potential richness for endemic avifauna, associated with higher altitude areas of around 450m to 850m.

- The areas of greatest richness refer to deciduous and semi-deciduous seasonal forest, steppe vegetation, areas of ecological tension between steppe vegetation and seasonal forest and areas of open woodland savannah.

- The existing Conservation Units do not sufficiently represent the areas of greatest richness, demonstrating that there is a need for new Conservation Units to be created in these areas.

CHAPTER 6

Bibliographical references

ALBUQUERQUE, S.G. & BANDEIRA, G.R.L. (1995) Effect of thining and slashing on forage phytomass from a Caatinga of Petrolina, Per nambuco, Brazil. Pesquisa Agropecuária Brasileira 30: 885-891.

ALBUQUERQUE, U.P. et al. (2012) Caatinga Revisited: Ecology and Conservation of an Important Seasonal Dry Forest. The Scientific World Journal Volume, 2012.

ALVES, J.J.A. (2007) Geoecology of the caatinga in the semi-arid region of northeastern Brazil. Climatology and Landscape Studies 2(1): 58-71.

ALVES, J.J.A., ARAÚJO, M.A. & NASCIMENTO, S.S. (2009) Degradation of the caatinga: an ecogeographic investigation. Revista Caatin ga 22(3): 126-135.

AMARAL, A.C.A. et al. (2005) Nest dynamics of the Leary Macaw (Anodorhynchus leari Bonaparte, 1856) in Jeremoabo, Bahia. Or nithologia 1(1): 59-64.

AMBDATA - Environmental variables for species modelling. Available at: < http://www.dpi.inpe.br/Ambdata/index.php> Accessed on: 15 Sep. 2013.

BIODIVERSITAS (2000) Biodiversity of the Caatinga Biome. Caatinga Workshop . Available at <http://www.biodiversitas.org.br/caatinga/bdados/mapsintese.asp>. Accessed on: 20 Sep. 2014

BIRDLIFE INTERNATIONAL (2013) Anodorhynchus leari. In: IUCN 2013. IUCN Red List of Threatened Species. Version 2013.2. Available at: <www.iucnredlist.org>. Accessed on: 06 Mar 2014.

BIRDLIFE INTERNATIONAL (2013) Cyanopsitta spixii. In: IUCN 2013. IUCN Red List of Threatened Species. Version 2013.2. Available at:

<www.iucnredlist.org>. Accessed on: 06 Mar 2014.

BIRDLIFE INTERNATIONAL (2016) Celeus obrieni. The IUCN Red List of Threatened Species 2016. Accessed on: 20 March 2017.

CAMPOS, M.E. (2012) Scarabaeidae (Coleoptera) in the Bahian Caatinga: Distribution and research planning. 50f. Dissertation (Honours Degree in Earth and Environmental Sciences) - Universidade Esta dual de Feira de Santana, Bahia.

CANHOS, D.A.L., SOUZA, S. & CANHOS, V. P. (2001) The use of electronic networks in biodiversity. Available at <http://www.comciencia.br/reportagens/biodiversidade/bio17.htm> Accessed on: 13 May 2014.

CANHOS, V.P. (2003) Informatics for biodiversity: Standards, protocols and tools. Ciência e Cultura 5(2): 45-47.

CARPENTER, G., GILLISON, A.N. & WINTER, J. (1993) Domain: a flex ible modelling procedure for mapping potential distributions of plants and animals. Biodiversity and Conservation 2: 667-680.

CASTELLETTI, C.H.M. et al. (2004) How much of the Caatinga is left? A preliminary estimate. In: J.M.C. Silva, M. Tabarelli, M.T. Fon seca & L.V. Lins (eds.). Biodiversity of the Caatinga: priority areas and actions for conservation. Ministry of the Environment, Brasília, pp. 91-100.

CLEERE, N. (1999) Family Caprimulgidae. In: DEL HOYO, J. ELLIOT, A., & CHRISTIE, D. A. Handbook of the Birds of the World. Vol. 5. Barn-owls to Hummingbirds. Lynx Edicions, Barcelona.

COLLAR, N.J. (1997) Family Psittacidae In: DEL HOYO, J. ELLIOT, A., & SARGATAL, J. (Eds). Handbook of the Birds of the World. Vol. 4. Sandgrousse to Cuckoos. Lynx Edicions, Barcelona.

DE MARCO JR, P. & SIQUEIRA, M.F. (2009) How to determine the potential distribution of species under a conservation approach? Megadiversity 5(1-2): 65-76.

DOS ANJOS, L. et al. (2009) Family Corvidae. In: DEL HOYO, J. EL LI OT, A., & CHRISTIE, D.A. (Eds). Handbook of the Birds of the World. Vol. 14. Bush-shrikes to Old World Sparrows. Lynx Edi cions, Bar celona.

FARBER, O. & KADMON, R. (2003) Assessment of alternative ap proaches for bioclimatic modelling with special emphasis on the Mahalanobis distance. Ecological Modelling 160(1): 115-130.

FERNANDES, J.D. & MEDEIROS, A.J.D. (2009) Desertification in the North: an approach to the phenomenon in Rio Grande Norte. Holos, Year 25, 3: 147-161.

FERRAZ, K.M.P.M.B. et al. (2012) Environmental suitability of a highly fragmented and heterogeneous landscape for forest bird species in south-eastern Brazil. Environmental Conservation 39 (4): 316324.

FIELDING, A.H. & BELL, J.F. (1997) A review of methods for the as sessment of prediction errors in conservation presence/absence models. Environmental Conservation 24 (1): 38-49.

FITZPATRICK, J.W. et al. (2004) Family Tyrannidae. In: DEL HOYO, J. ELLIOT, A., & CHRISTIE, D. A. Handbook of the Birds of the World. Vol. 9. Cotingas to Pipits and Wagtails. Lynx Edicions, Bar celona.

FORSHAW, J. M. (1978) Parrots of the world. Second edition. Lans downe Editions, Melbourne.

FORSHAW, J.M., & COOPER, W.T. (1989). Parrots of the world. Lon don: Blandford.

FRAGA, R.M. (2011) Family Icteridae (New world blackbirds) In: DEL HOYO, J. ELLIOT, A., & CHRISTIE, D. A. (Eds). Handbook of the Birds of the World.

Vol. 16. Tanagers to New World Blackbirds. Lynx Edicions, Barcelona.

GIANNINI, T.C. et al. (2012) Current challenges in predictive modelling of species distribution. Rodriguésia 63(3): 733-749.

GUEDES, T.B., NOGUEIRA, C. & MARQUES, O.A.V. (2014) Diversity, natural history, and geographic distribution of snakes in the Caatinga, Northeastern Brazil. Zootaxa 3863 (1): 1-93.

GUISAN, A. & TUILLER, W. (2005) Predicting species distribution: Offering more than simple habitat models. Ecology Letters 8: 993-1009.

GUISAN, A. & ZIMMERMANN, N.E. (2000) Predictive habitat distribution models in ecology. Ecological Modelling 135: 147-186.

HELLMAYR, C.E. (1929) A contribution to the ornithology of Northeast Brazil. Field Museum of Natural History - Zoology Series 12(18): 235-500.

HERNANDEZ, P.A. et al. (2006) The effect of sample size and species characteristics on performance of different species distribution modelling methods. Ecography 29: 773-785.

HIJMANS, R.J. et al. (2005) Very high resolution interpolated climate surfaces for global land areas. International Journal of Climatology 25: 1965-1978.

HIJMANS, R.J., & GRAHAM, C.H. (2006) The ability of climate envelope models to predict the effect of climate change on species distributions. Global Change Biology 12(12): 2272-2281.

HILTY, S. (2011) Scarlet-throated Tanager *(Compsothraupis loricata) In:* del Hoyo, J. et *al.* (eds.). Handbook of the Birds of the World Alive. Lynx Edicions, Barcelona. Available at: <http://www.hbw.com/node/61578> Accessed on: 21 Aug 2014.

IBGE - Brazilian Institute of Geography and Statistics (2007) Physical Map of Brazil. Available at < http://mapas.ibge.gov.br/fisicos/brasil> Accessed on: 10 Sep.

2014.

IBGE - Brazilian Institute of Geography and Statistics (2009) Continental Morphology and the Bottom of the Oceans. Available at <http://mapas.ibge.gov.br/fisicos/brasil> Accessed on: 10 Sep. 2014.

IBGE - Brazilian Institute of Geography and Statistics (2010) Population census 1940-2010. Available at:< http://seriesestatisticas.ibge.gov.br/series.aspx?vcodigo=POP122 > Accessed on: 31 Oct. 2014.

IBGE - Brazilian Institute of Geography and Statistics (2014) Brazil in a nutshell. Available at :< http://brasilemsintese.ibge.gov.br/territorio> Accessed: 23 May 2013.

ICMBio - Chico Mendes Institute (2011) Executive summary of the national action plan for the conservation of caatinga birds. Available at: < http://www.icmbio.gov.br/portal/biodiversidade/fauna- brasileira/plano-de-acao/866-plano-de-acao-nacional-para- conservacao-das-aves-da-caatinga. html> Accessed on: 20 May 2013.

IUCN (2016) The IUCN Red List of Threatened Species. Version 2013-2. Available at <http://www.iucnredlist.org>. Accessed on: 07 March 2014.

JARAMILLO, A. (2011) Red-cowled Cardinal (Paroaria dominicana). In: DEL HOYO, J. et al. (eds.). Handbook of the Birds of the World Alive. Lynx Edicions, Barcelona. Available at: <http://www.hbw.com/node/62100> Accessed on: 21 Aug 2014.

JARAMILLO, A. (2011) San Francisco Sparrow (Arremon franciscanus). In: DEL HOYO, J., et al. (eds.) Handbook of the Birds of the World Alive. Lynx Edicions, Barcelona. Available at: <http://www.hbw.com/node/61960> Accessed on: 21 Aug. 2014.

JARAMILLO, A. (2011) White-throated Seedeater *(Sporophila albogu laris). In:* DEL

HOYO, J. et *al.* (eds.) Handbook of the Birds of the

World Alive. Lynx Edicions, Barcelona. Available at: <http://www.hbw.com/node/62122> Accessed on: 21 Aug 2014.

KRABBE, N.K. & SCHULENBERG, T.S. (2003) Family Rhinococrypti dae. In: Handbook of the Birds of the World. Vol. 8. Broadbills to Tapaculos. Lynx Edicions, Barcelona. DEL HOYO, J. ELLIOT, A., & CHRISTIE, D. A.

LEAL, I.R. et al. (2005) Changing the course of biodiversity conservation in the Caatinga of Northeast Brazil. Megadiversity 1 (1): 139-146.

LEAL, I.R., TABARELLI, M. & SILVA, J.M.C. (2003) Ecology and Conservation of the Caatinga: An Introduction to the Challenge. In: Leal, I.R. et al. Ecologia e Conservação da Caatinga. Editora Universitária UFPE, p. XIII-XVI.

LEITE, L.O. (2006) Analysis of endemism, geographical variation and potential distribution of Cerrado endemic bird species. Brasília, DF. 181 f. Thesis (Doctorate in Animal Biology) - University of Brasília Institute of Biological Sciences. 2006.

LIMA-RIBEIRO, M.S. & DINIZ-FILHO, J.A.F. (2012) Modelling the geographical distribution of species in the past. Revista brasileira de paleontologia 15(3): 371-385.

LUIZ, E.R. (2010) Conservation of the Gravel-breasted Sandpiper Rhopornis ardesiacus (Wied 1831) (Aves: Thamnophilidae): Geographic distribution, foraging strategy, density and population estimate. Master's thesis.

MACHADO, C.G. (2006) The Birds of the Semi-Arid Region of Bahia. In: QUEI ROZ, L. P., RAPINI, A. & GIULIETTI, A. M. Rumo ao Amplo Co nhecimento da Biodiversidade do Semi-árido Brasileiro. Ministry of Science and Technology, Brasília.

MARANTZ, C.A. et al. (2003) Family Dendrocolaptidae. In: Handbook of the Birds of the World. Vol. 8. Broadbills to Tapaculos. Lynx Edi cions, Barcelona. DEL HOYO, J. ELLIOT, A., & CHRISTIE, D. A.

MARINI, M.A. & LOPES, L.E. (2005) New southern limit in the geographic distribution of Sakesphorus cristatus (Thamnophilidae). Ararajuba 13(1): 105-106.

MARINI, M.A. et al. (2009) Major current and future gaps of Brazilian re serves to protect Neotropical savanna birds. Biological Conserva tion 142: 3039-3050.

MARINI, M.A. et al. (2010) Predicting the occurrence of rare Brazilian birds with species distribution models. Journal of Ornithology 151:857-866.

METZ, C.E. (1986) ROC methodology in radiologic imaging. Investiga tional Radiology 21:720-733.

MMA - Ministry of the Environment (1999) First national report for the convention on Biological Diversity. Brazil. Secretariat for Biodiversity and Forests, MMA, Brasília.

MMA - Ministry of the Environment (2002) Evaluation and identification of priority areas and actions for the conservation, sustainable use and sharing of the benefits of biodiversity in Brazilian biomes. MMA/SBF.

MMA - Ministry of the Environment (2004) Conservation Units: priority areas and actions for the conservation of the Caatinga. Brasília, DF.

MMA - Ministry of the Environment (2007) Priority Areas for the Conservation, Sustainable Use and Benefit Sharing of Brazilian Bio Diversity: MMA Ordinance No. 09.

MMA - Ministry of the Environment (2014) Caatinga. Available at: <http://www.mma.gov.br/biomas/caatinga> Accessed on: 23 May 2013.

MORATO, R.G. et al. (2014) Identification of priority conservation areas and potential corridors for jaguars in the Caatinga biome, Brazil. Plos one 9(4): 1-11.

MOURA, D.C. (2003) Richness and abundance of bees in different stages of caatinga degradation as environmental indicators around the xingó hydroelectric power station. 2003. Recife, Pernambuco. 96 f. Dissertation (Master's Degree in Environmental Management and Policies) - Federal University of Pernambuco.

MUNOZ, M.E.S. et *al.* (2009) OpenModeller: a *generic* approach to spe cies' potential distribution modelling. GeoInformatica 15(1): 111135.

NETO, O.L. (2013) Biodiversity Conservation and Historical Biogeography. In: CARVALHO, C.J.B. & ALMEIDA, E.A.B. (Eds) Biogeography of South America: Patterns and Processes. São Paulo: Editora Rocca, pp. 162-172.

NIMER, E. (1972) Climatology of the Northeast Region of Brazil. Introduction to dynamic climatology. Brazilian Journal of Geography 34(2): 351.

OLIVEIRA, H. R. & CASSEMIRO, F.A.S. (2013) Potential effects of future climate change on the distribution of a Caatinga anuran Rhinella granulosa (Anura, Bufonidae). Inheringia, Série Zoologia, Porto Alegre, 103(3): 272-279.

OLIVEIRA, R.G. & MEDEIROS, W.E. (2012) Evidences of buried loads in the base of the crust of Borborema Plateau (NE Brazil) from Bouguer admittance estimates. Journal of South American Earth Sciences 37: 60-76.

OLMOS, F & ALBANO, C. (2012) The birds of the Serra da Capivara National Park region (Piauí, Brazil). Brazilian Journal of Ornithology 20(3): 173-187.

OLMOS, F., SILVA, W.A.G. & ALBANO, C.G. (2005) Birds in eight areas of caatinga in southern Ceará and western Pernambuco, northeastern Brazil: Composition, richness and similarity. Papéis Avulsos de Zo ologia 45(14): 179-199.

PACHECO, J.F. (2004) The birds of the Caatinga: a historical analysis of knowledge. In: SILVA, J.M.C., TABARELLI M., FONSECA M. T. & LINS, L.V. (Orgs.). Biodiversity of the Caatinga: priority areas and actions for conservation. Brasília: MMA/UFPE, pp. 189-250.

PEARCE, J. & FERRIER, S. (2000) Evaluating the predictive performance of habitat models developed using logistic regression. Eco logical Modelling 133: 225-245.

PEARSON, R.G. (2007) Species' Distribution Modelling for Conservation Educators and Practitioners. Synthesis. American Museum of Nat ural History 1: 1-50

PHILLIPS, S.J., ANDERSON, R.P. & SCHAPIRE, R.E. (2006) Maximum entropy modelling of species geographic distributions. Ecological Modelling 190: 231-259.

PRADO, D.E. (2003) The caatingas of South America. In: LEAL, I.R.; TABARELLI, M.; SILVA, J.M.C. (Eds.) Ecologia e conservação da Caatinga. Recife: Editora Universitária da UFPE, pp.3-74.

REMSEN, J.V. (2003) Family Furnariidae. In: Handbook of the Birds of the World. Vol. 8. Broadbills to Tapaculos. Lynx Edicions, Barcelo na. DEL HOYO, J. ELLIOT, A., & CHRISTIE, D. A.

RIBEIRO, V. (2014) Biogeography of South American savannah birds. 2014. Brasília, DF. 99 f. Dissertation (Master's in Ecology) - University of Brasília Institute of Biological Sciences.

ROOS, A.L. et al. (2006) Avifauna of the Sobradinho Lake region: composition, richness and biology. Ornithologia 1(2): 135-160.

SAMPAIO, E.V.S.B. (1995) Overview of the Brazilian caatinga. In: BULLOCK, S.H., MOONEY, H.A. & MEDINA, E. (eds.) Seasonal dry tropical forests. Cambridge University Press, Cambridge, pp. 35-63.

SANTOS NETO, J.R. & CAMANDAROBA, M. (2007) Expansion of the area of occurrence of the Lear's macaw Anodorhynchus leari (Bona parte 1856). Ornithologia 2(1): 63-64.

SANTOS, D.P.M. (2004) Bird communities in two physiognomies of Caatinga vegetation in the state of Piauí, Brazil. Ararajuba 12(2): 113-123.

SANTOS, E.B.G., DANTAS, I.M. & RAJÃO, H. (2012) The contribution of WikiAves to locality bases in biogeographic studies: a case study with the genus Drymophila (Thamnophilidae). In: XIX CONGRESSO BRASILEIRO DE ORNITOLOGIA, Alagoas, p. 319. Electronic Proceedings, Maceió, AL. Bird conservation in Brazil: The northeast discusses the current situation and future prospects   : <http://www.ararajuba.org.br/sbo/cbo/cbo.htm> Accessed on: 1 Feb. 2014

SCHUCHMANNM, K.L. (1999) Family Trochlidae. In: DEL HOYO, J. EL LIOT, A., & SARGATAL, J. Handbook of the Birds of the World. Vol. 5. Barn-owls to Hummingbirds. Lynx Edicions, Barcelona.

SHORT, L. L. (1982) Picumnidae In: Woodpeckers of the world. Dela ware Museum of Natural History.

SIBLEY, C.G. & MONROE JR, B.L. (1990) Distribution and Taxonomy of Birds or the World. Yale University Press.

SICK, H. (1985) Ornitologia brasileira, uma introdução. Brasília. University of Brasilia. Two volumes.

SICK, H. (1997) Ornitologia Brasileira. Revised and expanded edition by Jo sé Fernando Pacheco. Editora Nova Fronteira S. A.

SILVEIRA, L.F. & STRAUBE, F.C. (2008) Birds. In: MACHADO, A.B.M., DRUMMOND, G.M. & PAGLIA, A.P. Livro Vermelho da Fauna Brasileira Ameaçada de Extinção. Biodiversity 19. Brasília, DF.

SILVEIRA, L.F., & STRAUBE, F.C. (2008). Endangered birds in Brazil. Livro vermelho da fauna brasileira ameaçada de extinção, vol. 2, pp. 378-679.

SOBERÓN, J. & PEARSON, A.T. (2005) Interpretation of models of fun damental ecological niches and species' distributional areas. Biodiversity Informatics 2:

1-10.

SOUZA, E.A. et al. (2009) Expansion of the occurrence area of the Red-crested Hummingbird Augastes lumachella (Lesson , 1838) (Trochilidae). Ornithologia 3(2):145-148

SOUZA, T.V. et al. (2011) Redistribution of Threatened and Endemic Atlantic Forest Birds Under Climate Change. Natureza & Com servação 9(2):214-218.

SPECIESLINK (2012) Electronic Database. Available at <http://splink.cria.org. br/>. Accessed on 15 Nov. 2013

STOTZ, D.F. (Eds) (1996) Neotropical birds: Ecology and conservation. University of Chicago Press.

TABARELLI, M. & VICENTE, A. (2004) Knowledge about Caatinga woody plants: geographical and ecological gaps. In: Silva, J.M.C.; Tabarelli, M., Fonseca, M. & Lins, L. (Eds.). Biodiversity of the Caatinga: priority areas and actions for conservation. Ministry of the Environment, Brasília, pp. 101-110.

THE NATURE CONSERVANCY DO BRASIL & ASSOCIAÇÃO CAA TINGA (2000) Evaluation and identification of priority actions for the conservation, sustainable use and benefit sharing of the biodiversity of the Caatinga biome: CONSERVATION UNITS IN THE CAATINGA. Petrolina, pp. 1-9.

VASCONCELOS, M.F, LOPES, L.E., MACHADO, C.G. & RODRIGUES, M. (2008) The birds of the rupestrian grasslands of the Espinhaço Chain: diversity, endemism and conservation. Megadiversity 4(1-2): 197-217.

VELLOSO, A.L., SAMPAIO, E.V.S.B. & PAREYN, F.G.C. (2001) Ecoregions proposed for the Caatinga biome. Results of the Caatinga Ecoregional Planning Seminar/Aldeia-PE. 20 to 30 November 2001.

VIEIRA, C.S. (2007) A representatividade das unidades de conservação do bioma Mata Atlântica da Bahia na conservação da avifauna ameaçada. Ilheus, Bahia.

83 f. Dissertation (Master's in Regional Development and Environment) - State University of Santa Cruz.

WHITNEY, B.M. et al. (2000) Systematic revision and biogeography of the Herpsilochmus pileatus complex, with description of a new species from northeastern Brazil. The Auk 117(4): 869-891.

WIKIAVES (2013) Available at <http://www.wikiaves.com.br/> Accessed on: 15 May 2014.

WINKLER H. & CHRISTIE, A. (2002) Family Picidae. In: DEL HOYO, J. ELLIOT, A., & CHRISTIE, D. A. Handbook of the Birds of the

World. Vol. 7. Jacamars to Woodpeckers. Lynx Edicions, Barcelo na.

ZANELLA, F.C.V. (2013) Evolution of the Biota of the Diagonal Dry Open Formations of South America. In: CARVALHO, C. J. B. & ALEMEIDA, E.A.B. (Eds) Biogeography of South America: Patterns and processes. São Paulo: Editora Rocca, pp. 198-220.

ZIMMER, K.J. & ISLER, M.L. (2003) Family Thamnophilidae In: DEL HOYO, J. ELLIOT, A., & CHRISTIE, D. A. Handbook of the Birds of the World. Vol. 8. Broadbills to Tapaculos. Lynx Edicions, Barcelo na.

ZIMMERMANN, N.E. et al. (2010) New trends in species distribution modelling. Ecography 33(6): 985-989.

CHAPTER 7

Appendices

Appendix A: Values of the most correlated environmental variables (above 0.7). Caption: BIO 1 - Average annual temperature; BIO 2 - Average diurnal variation; BIO 3 - Isothermality; BIO 4 - Seasonal temperature; BIO 5 - Maximum temperature of the hottest month; BIO 6 - Minimum temperature of the coldest month; BIO 7 - Annual temperature range; BIO 8 - Average temperature of the wettest quarter; BIO 9 - Average temperature of the driest quarter; BIO 10 - Average temperature of the hottest quarter; BIO 11 - Average temperature of the coldest quarter; BIO 12 - Annual precipitation; BIO 13 - Precipitation of the wettest month; BIO 14 - Precipitation of the driest month; BIO 15 - Seasonal precipitation; BIO 16 - Precipitation of the wettest quarter; BIO 17 - Precipitation of the driest quarter; BIO 18 - Precipitation of the warmest quarter; BIO 19 - Precipitation of the coldest quarter.

Variables	Correlation value
BIO 10 and BIO 11	0.93228827
BIO 9 and BIO 6	0.94731183
BIO 14 and BIO 17	0.993716205
BIO 18 and BIO 4	0.7158304
BIO 7 and BIO 2	0.9210076
BIO 8 and BIO 5	0.86127805
BIO 13 and BIO 16	0.987164027
BIO 1 and BIO 9	0.94433015
BIO 11 and BIO 1	0.98333573
BIO 16 and BIO 12	0.87979127
BIO 11 and BIO 5	0.84241522

Appendix B: AUC values for all algorithms.

Species	Algorithm	AUC
Penelope jacucaca	MaxEnt	0.992
	Bioclim	0.983
	Mahalanobis	0.658
	SVM	0.985
Anodorhynchus leari	MaxEnt	0.999
	Bioclim	0.999
	Mahalanobis	0.501
	SVM	0.997
Eupsittula rum cactus	MaxEnt	0.986
	Bioclim	0.973
	Mahalanobis	0.738
	SVM	0.986
Hydropsalis hirundinacea	MaxEnt	0.994
	Bioclim	0.988
	Mahalanobis	0.706
	SVM	0.991
Anopetia gounellei	MaxEnt	0.994
	Bioclim	0.990
	Mahalanobis	0.681
	SVM	0.988
Augastes lumachella	MaxEnt	0.998
	Mahalanobis	0.818
	Bioclim	0.998
	SVM	0.995
Picumnus pygmaeus	MaxEnt	0.991
	Bioclim	0.968
	Mahalanobis	0.719
	SVM	0.984
Picumnus fulvescens	MaxEnt	0.998
	Bioclim	0.998
	Mahalanobis	0.769
	SVM	0.995
Picumnus limae	MaxEnt	0.997
	Bioclim	0.997
	Mahalanobis	0.732
	SVM	0.997
Myrmorchilus strígilatus	MaxEnt	0.991
	Bioclim	0.984
	Mahalanobis	0.682
	SVM	0.983
Formicivora iheríngi	MaxEnt	0.999
	Bioclim	0.999

Species	Method	Value
	Mahalanobis	0.503
	SVM	0.996
Herpsilochmus sellowi	MaxEnt	0.990
	Bioclim	0.979
	Mahalanobis	0.663
	SVM	0.986
Herpsilochmus pectoralis	MaxEnt	0.996
	Bioclim	0.994
	Mahalanobis	0.735
	SVM	0.990
Sakesphorus cristatus	MaxEnt	0.993
	Bioclim	0.983
	Mahalanobis	0.680
	SVM	0.983
Thamnophilus capistratus	MaxEnt	0.991
	Bioclim	0.983
	Mahalanobis	0.756
	SVM	0.988
Hylopezus ochroleucus	MaxEnt	0.997
	Bioclim	0.995
	Mahalanobis	0.636
	SVM	0.991
Xiphocolaptes falcirostris	MaxEnt	0.997
	Bioclim	0.998
	Mahalanobis	0.827
	SVM	0.989
Megaxenops parnaguae	MaxEnt	0.995
	Bioclim	0.995
	Mahalanobis	0.623
	SVM	0.989
Christian pseudosexuality	MaxEnt	0.987
	Bioclim	0.974
	Mahalanobis	0.748
	SVM	0.994
Synallaxis hellmayri	MaxEnt	0.996
	Bioclim	0.997
	Mahalanobis	0.681
	SVM	0.993
Hemitriccus mirandae	MaxEnt	0.999
	Bioclim	0.998
	Mahalanobis	0.747
	SVM	0.998
Stigmatura budytoides	MaxEnt	0.990
	Bioclim	0.997
	Mahalanobis	0.718

	SVM	0.992
Paroaria dominicana	MaxEnt	0.973
	Bioclim	0.953
	Mahalanobis	0.688
	SVM	0.980
Sporophila albogularis	MaxEnt	0.984
	Bioclim	0.963
	Mahalanobis	0.734
	SVM	0.986
Arremon franciscanus	MaxEnt	0.999
	Bioclim	0.999
	Mahalanobis	0.501
	SVM	0.992
Agelaioides fringillarius	MaxEnt	0.989
	Bioclim	0.978
	Mahalanobis	0.767
	SVM	0.990
Cyanocorax cyanopogon	MaxEnt	0.979
	Bioclim	0.965
	Mahalanobis	0.720
	SVM	0.981
Icterus jamacaii	MaxEnt	0.974
	Bioclim	0.939
	Mahalanobis	0.723
	SVM	0.979
Compsothraupis loncata	MaxEnt	0.988
	Bioclim	0.975
	Mahalanobis	0.688
	SVM	0.985

Appendix C: Consensus maps of the potential distribution of the 29 species modelled.

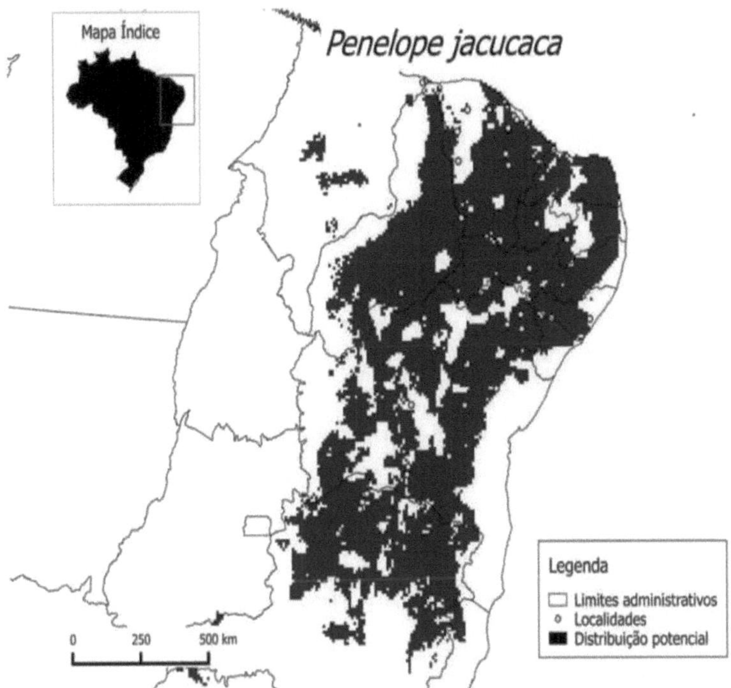

Penelope jacucaca: Distributed in north-eastern Brazil except along the coast. From Piauí, a portion of Ceará, Rio Grande do Norte, Paraíba, Pernambuco, Alagoas, Sergipe, down through central Bahia to northern Minas Gerais.

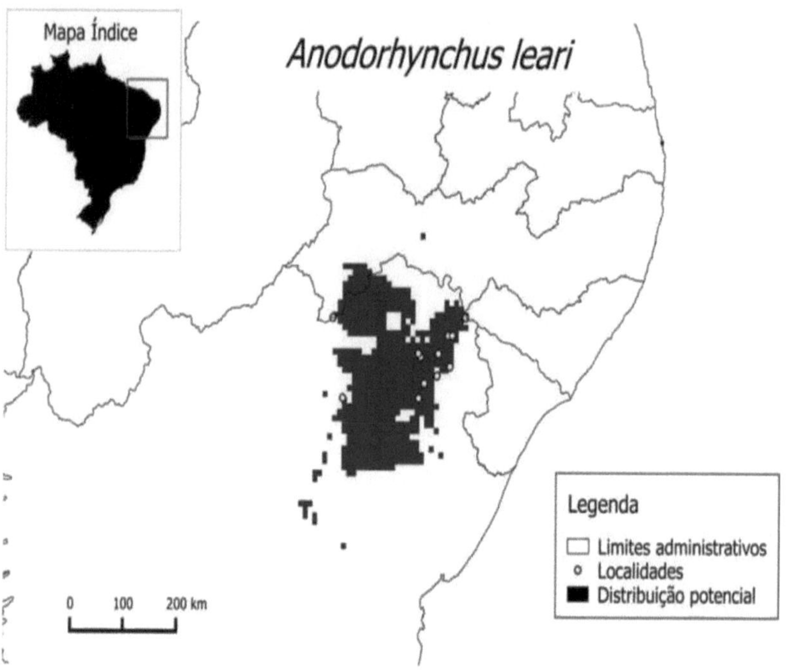

Anodorhynchus leari: It has a restricted distribution, occupying just one spot in the north of Bahia.

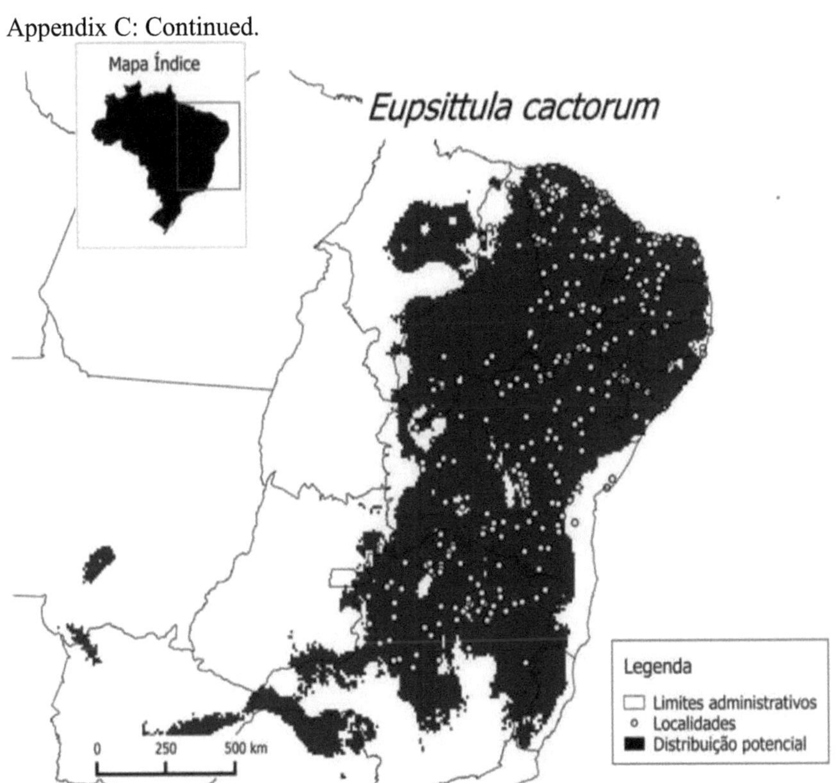

Eupsittula cactorum: It is distributed in the centre of Maranhão, from Piauí eastwards to Rio Grande do Norte, continuing southwards to the centre and Minas Gerais. It occupies the coast except for the state of Bahia.

Appendix C: Continued.

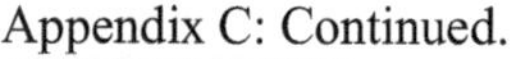

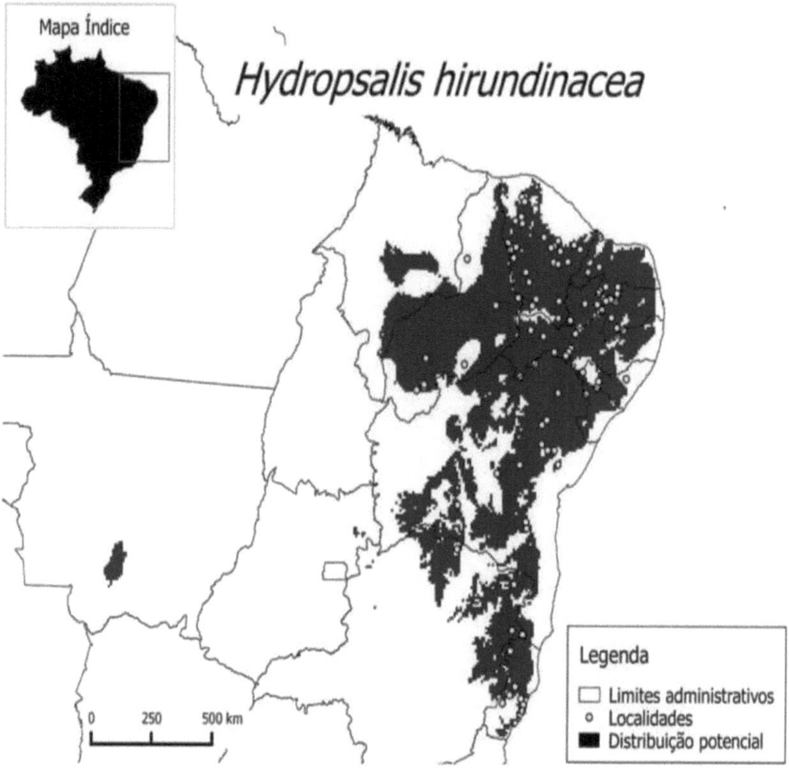

Hydropsalis hirundinacea: Distributed in north-eastern Brazil except along the coast. It covers the state of Piauí in an easterly direction up to Rio Grande do Norte, continuing in a southerly direction occupying the centre of Bahia, the north of Minas Gerais and the north of Espírito Santo.

Appendix C: Continued.

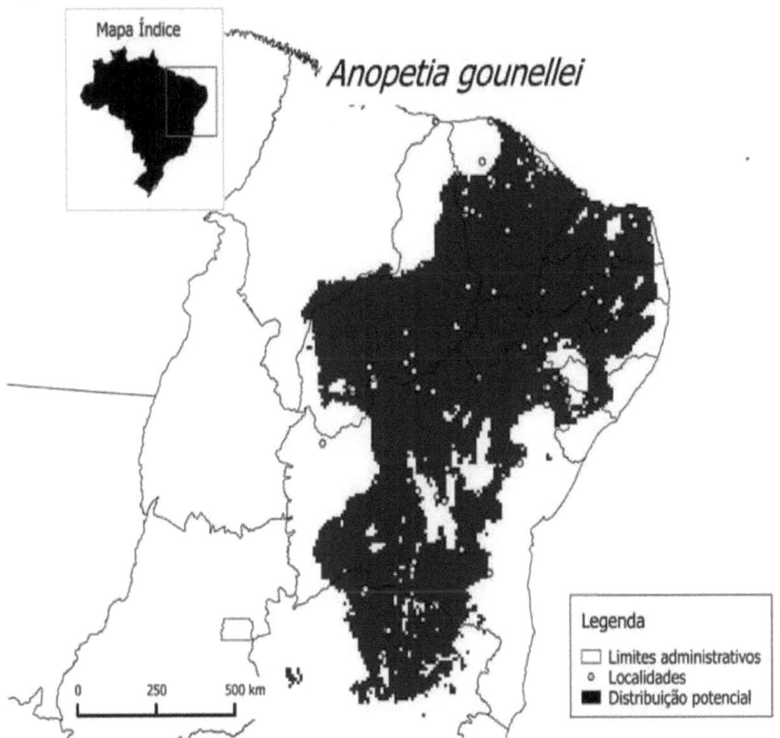

Anopetia gounellei: Occupies the east and south of Piauí, Ceará, Rio Grande do
Norte, Paraíba, Pernambuco, a strip in the centre of Alagoas, the west of Sergipe and
the centre of Bahia to the north of Minas Gerais.

Appendix C: Continued.

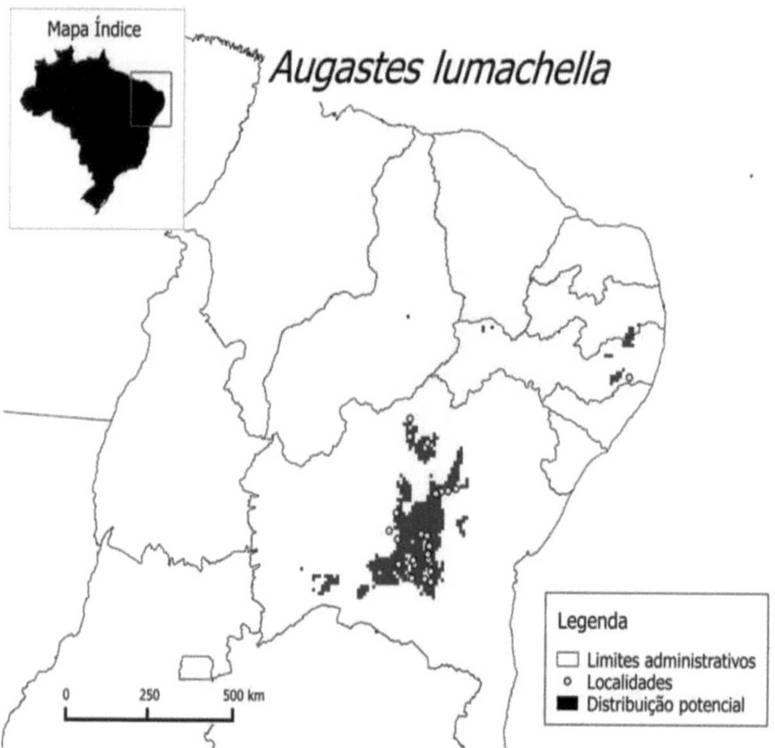

Augastes lumachella: Distributed in the centre of Bahia.

Appendix C: Continued.

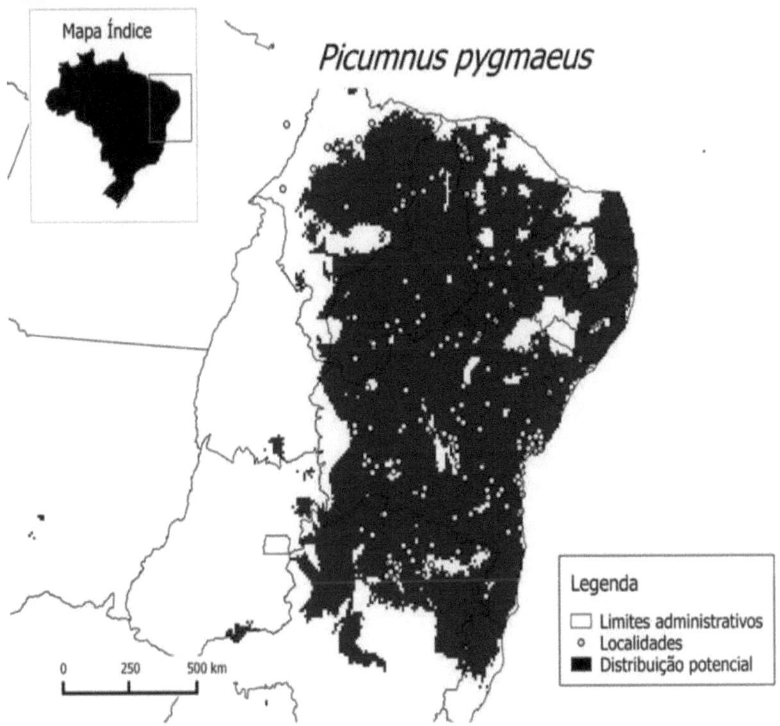

Picumnus pygmaeus: Distributed in eastern Maranhão, the state of Piauí, Ceará, except for the north, Rio Grande do Norte, Paraíba, Pernambuco, Alagoas, Sergipe, Bahia, northern Minas Gerais and northern Espirito Santo.

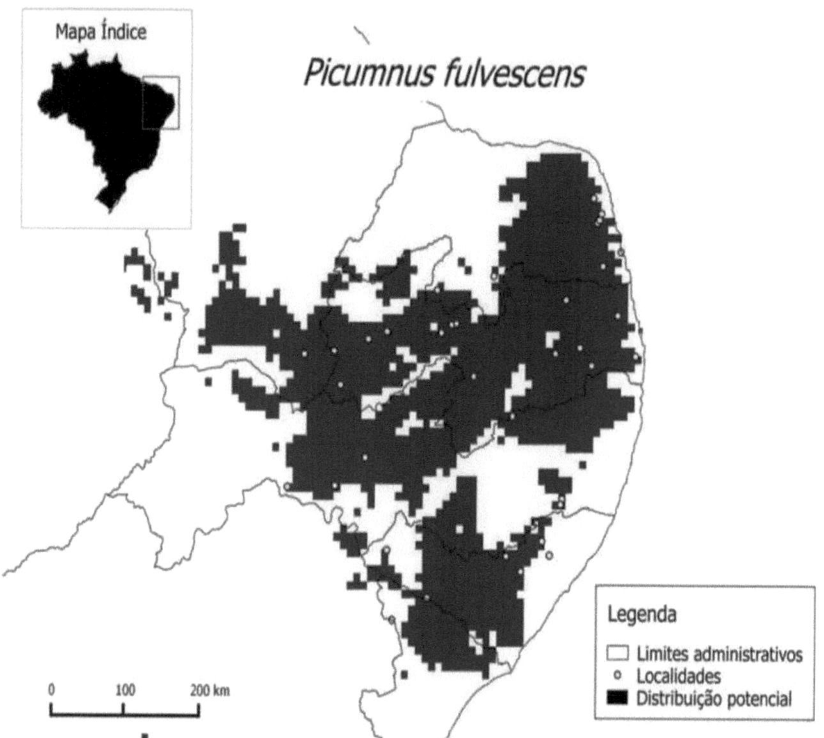

Picumnus fulvescens: Distributes to the east of Rio Grande do Norte, in the state of Paraíba, the extreme south of Ceará, the centre of Pernambuco, including the northeast, as well as the centre of Alagoas and the north of Sergipe.

Appendix C: Continued.

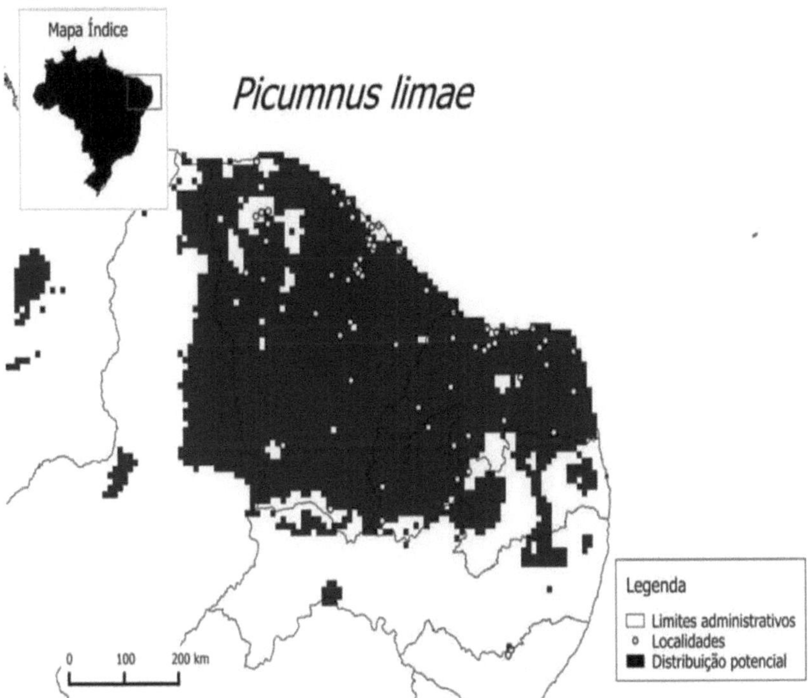

Picumnus limae: Distributed in the far east of Piauí, in the states of Ceará, Rio Grande do Norte and Paraíba.

63

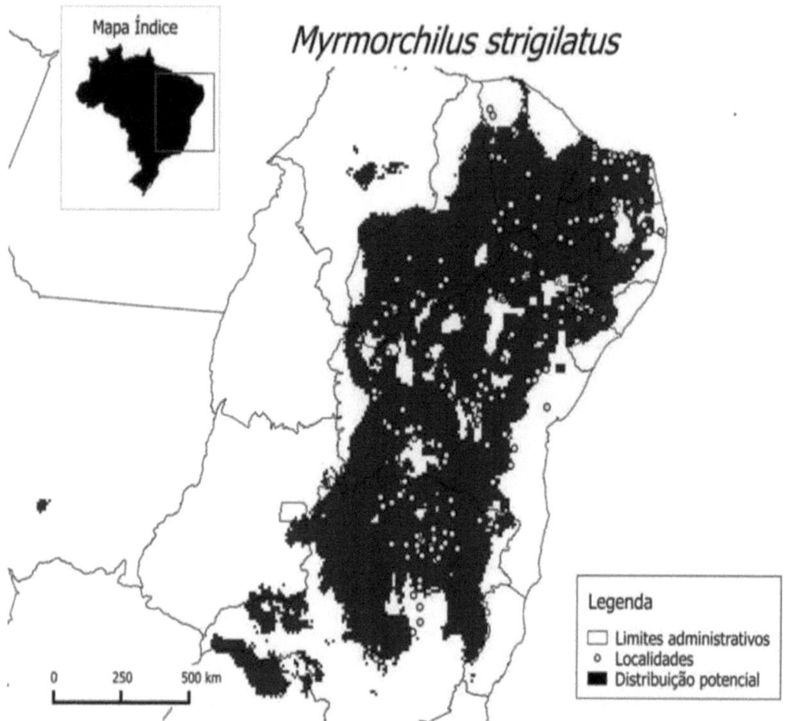

Myrmorchilus strigilatus: Distributed in north-eastern Brazil, except along the coast. It covers the east and south of Piauí, Ceará, with the exception of the northern region, Rio Grande do Norte, Paraíba, Pernambuco, down through the centre of Bahia to the centre of Minas Gerais.

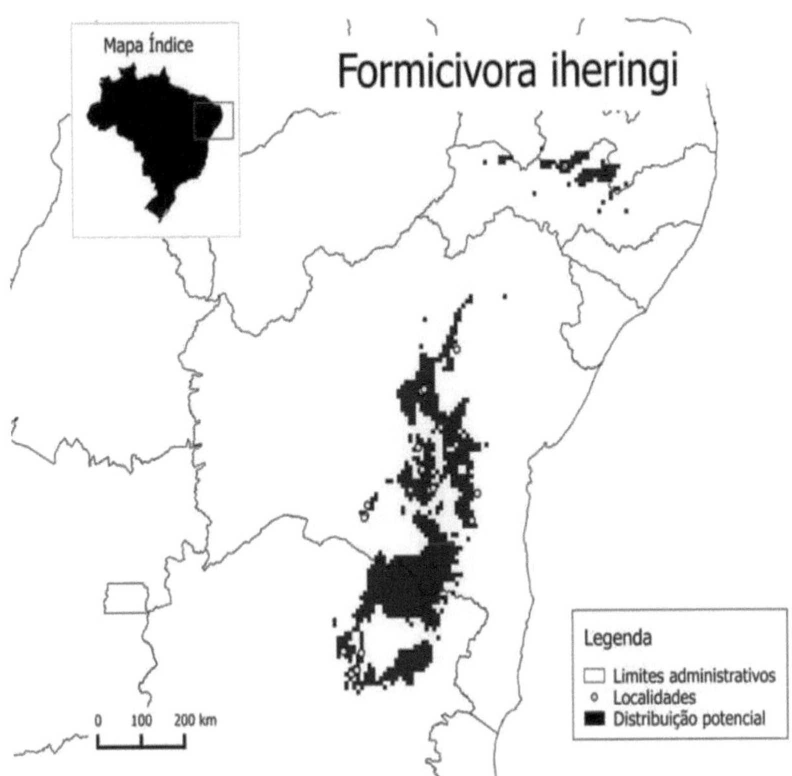

Formicivora iheringi: Distributed in the north of Minas Gerais with patches of distribution in the centre and south of Bahia, as well as small sites in the north of Pernambuco.

Appendix C: Continued.

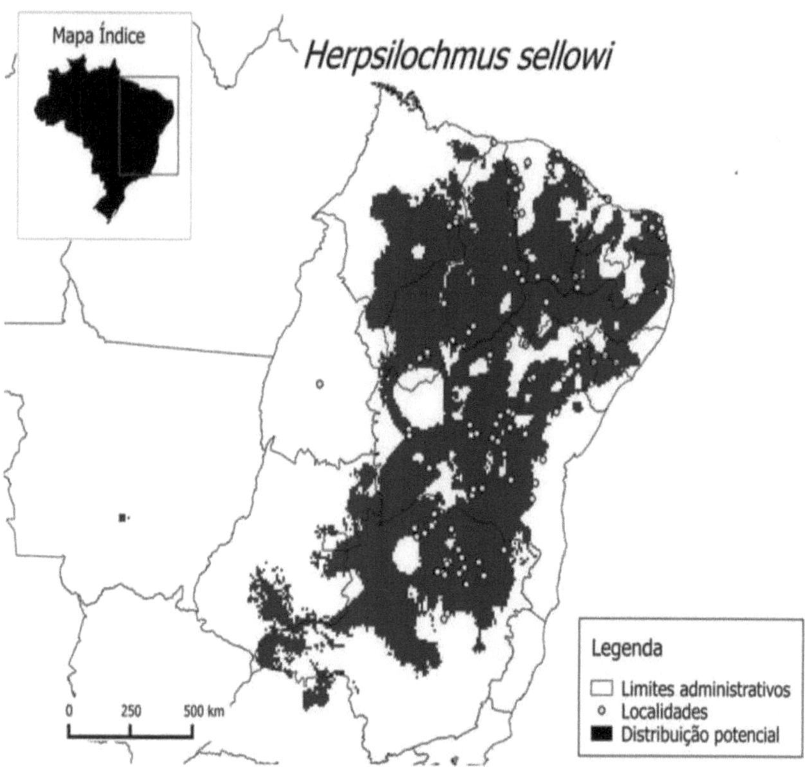

Herpsilochmus sellowi: Distributed in north-eastern Brazil except along the coast. It ranges from the south-east of Ceará to the east of Rio Grande do Norte and Paraíba, and southwards to the centre of Minas Gerais.

Appendix C: Continued.

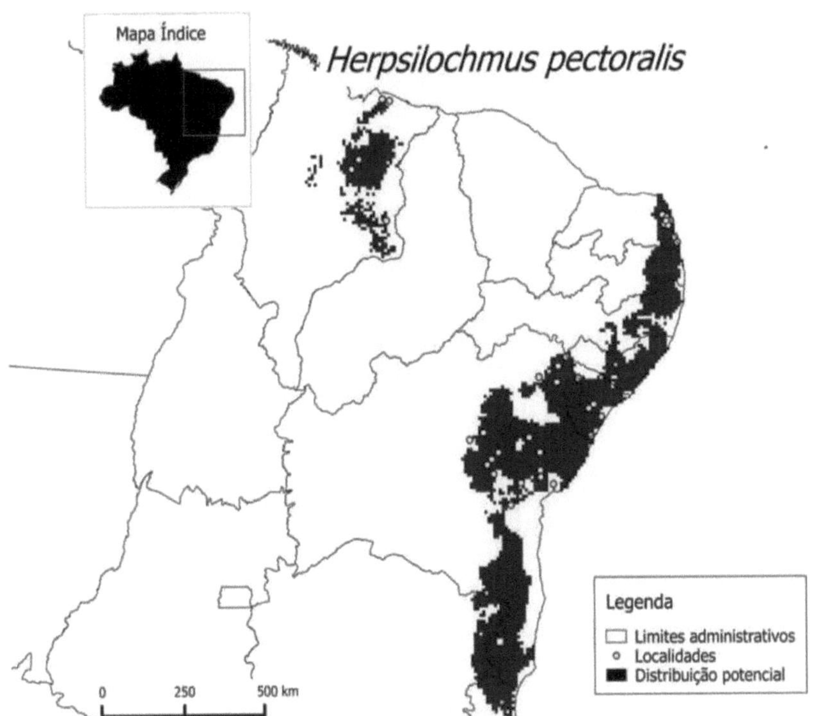

Herpsilochmus pectoralis: Distributed along the northeast coast of Brazil, including Rio Grande do Norte, Paraíba, Pernambuco, Alagoas, Sergipe, Bahia and northern Espírito Santo. In addition to an isolated spot in eastern Maranhão.

Sakesphorus cristatus: Distributed in the north-east of Brazil, except along the coast, in the south-east of Maranhão, the whole of Piauí, the south of Ceará, with a strip extending to the north, Rio Grande do Norte, Paraíba, Pernambuco, Alagoas, Sergipe, Bahia, the north of Minas Gerais, the east of Goiás, as well as a strip extending to the north of Mato Grosso do Sul and the south of Mato Grosso.

Appendix C: Continued.

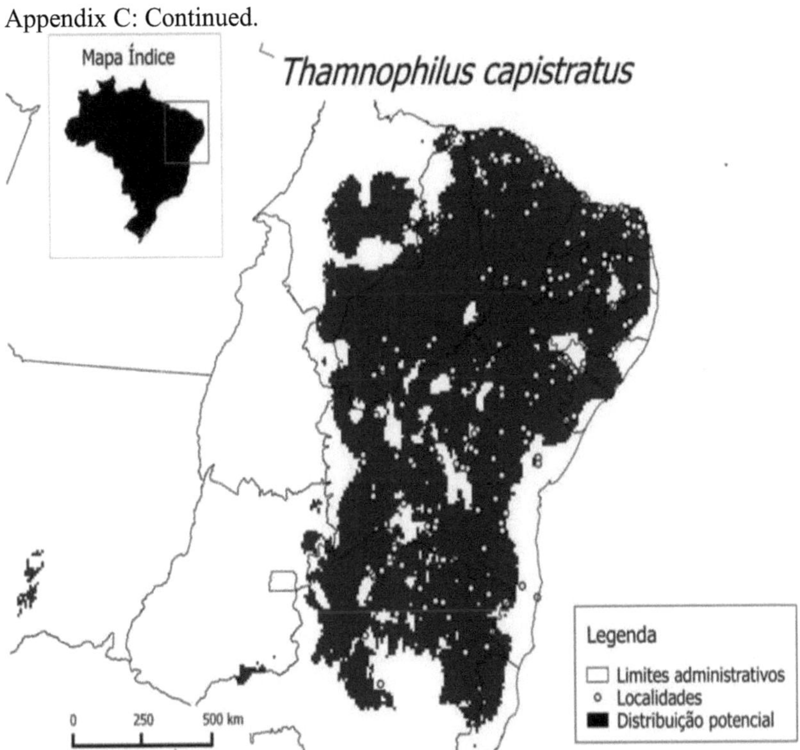

Thamnophilus capistratus: Distributed in north-eastern Brazil, except along the coast. It is suitable for the centre of Maranhão, the entire state of Piauí, Ceará, Rio Grande do Norte, Paraíba, Pernambuco, Alagoas, Sergipe, Bahia, northern Minas Gerais and western Espírito Santo.

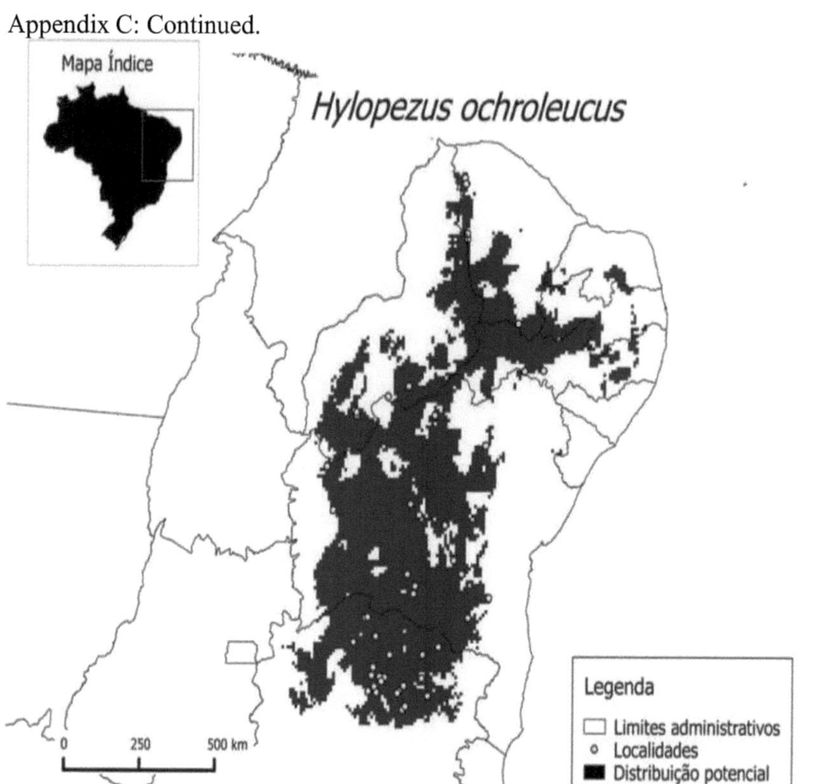

Hylopezus ochroleucus: Distributed in the far east and south of Piauí, south-west of Ceará, west of Pernambuco, down to the centre of Bahia and north of Minas Gerais.

Appendix C: Continued.

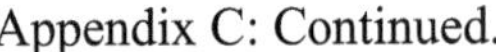

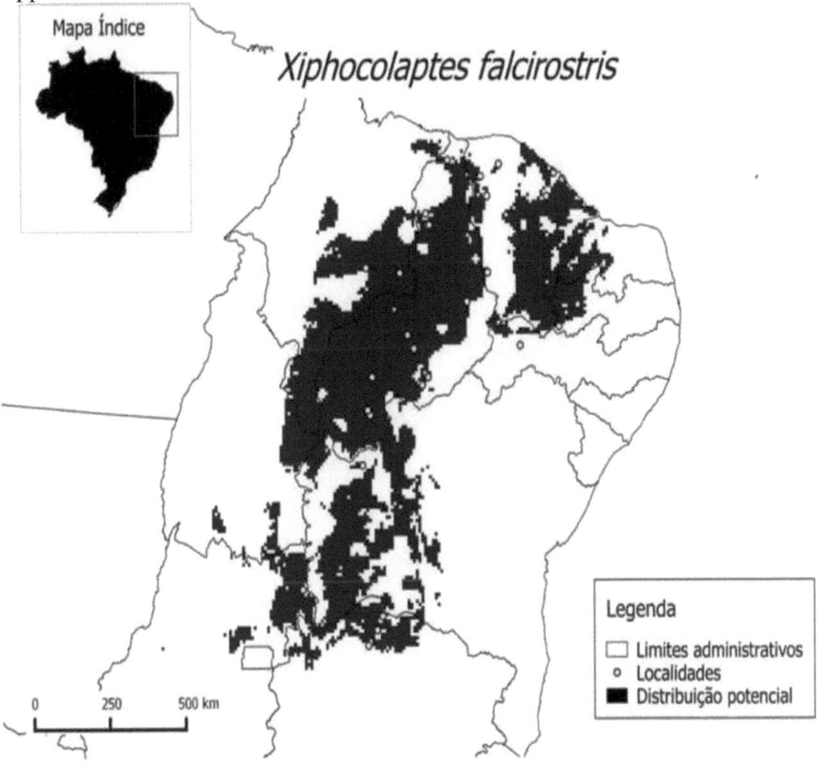

Xiphocolaptes falcirostris: Distributed in the south-east of Maranhão, west of Piauí, a patch in the east of Ceará with a patch down the west of Bahia to the extreme north of Goiás and Minas Gerais. There is also a patch in the far west of Rio Grande do Norte and Paraíba.

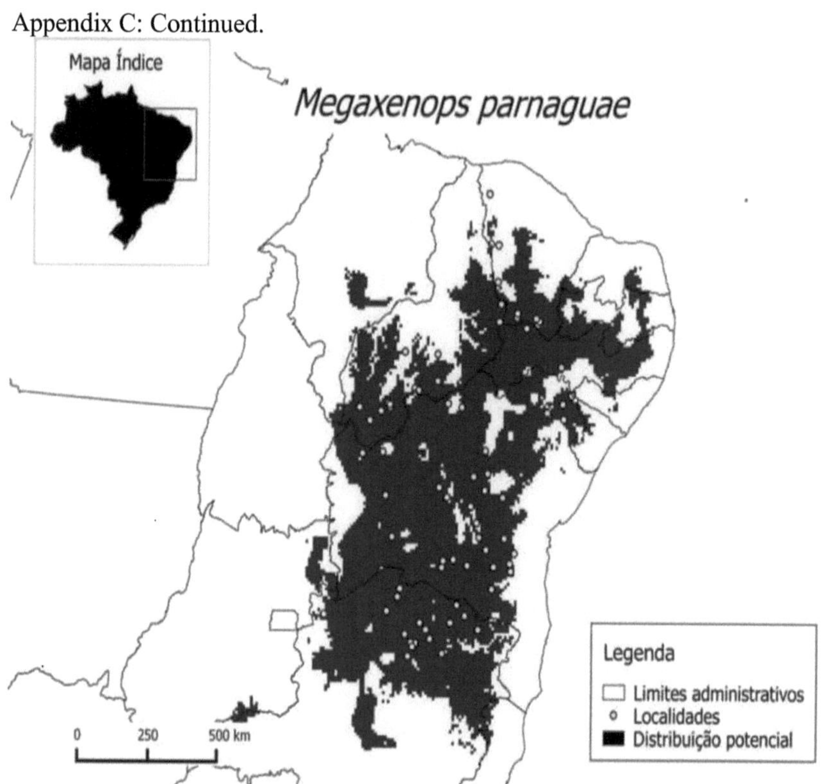

Megaxenops parnaguae: Distributed in the east and south of Piauí, south of Ceará, far west and south of Paraíba, west of Pernambuco, down to the centre of Bahia and north of Minas Gerais.

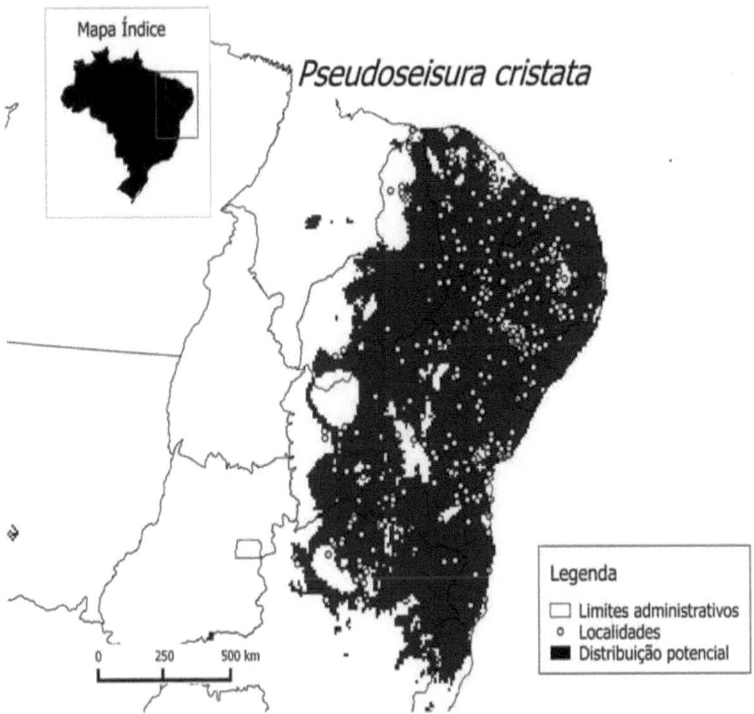

Pseudoseisura cristata: Distributed in eastern Piauí, Ceará, Rio Grande do Norte, Paraíba, Pernambuco, Alagoas, Sergipe, Bahia, northern Minas Gerais and Espírito Santos.

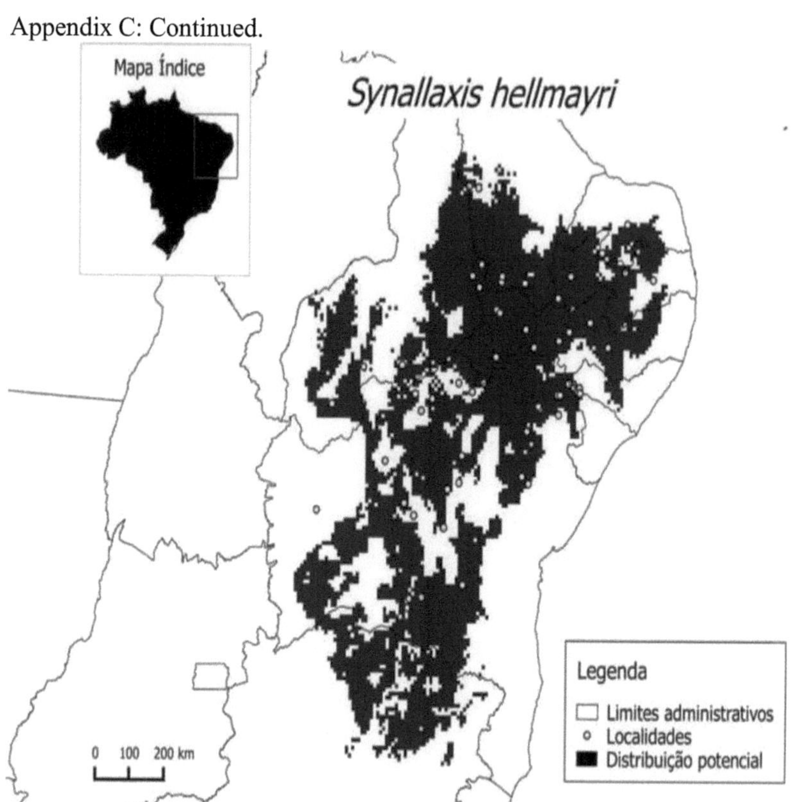

Synallaxis hellmayri: Distributes to the east of Piauí, south of Ceará, west of Paraíba, Pernambuco, except to the east, a strip extending to the centre of Bahia and the extreme north of Minas Gerais.

Appendix C: Continued.

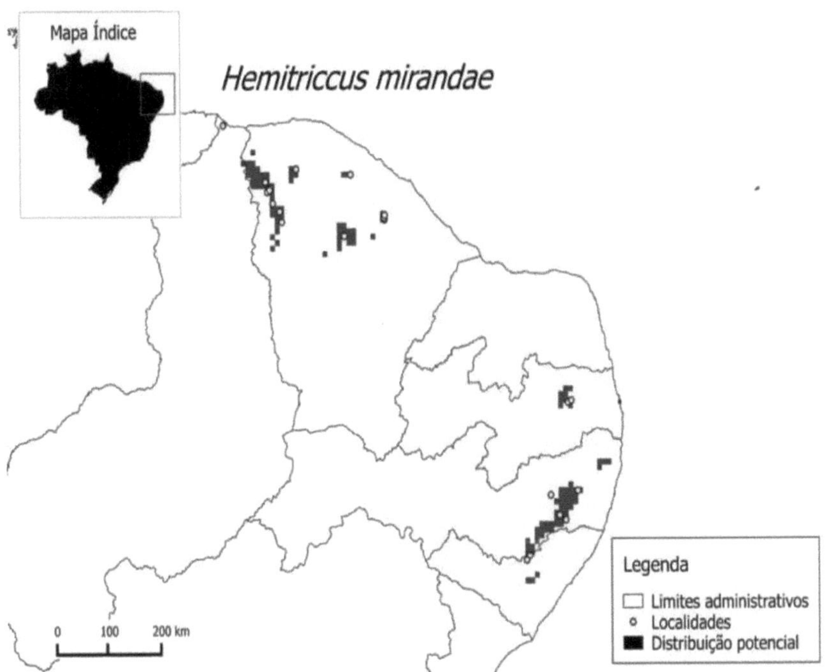

Hemitriccus mirandae: Isolated patches of distribution in the east of Pernambuco, Paraíba, Sergipe and Rio Grande do Norte and others in the north of Ceará and Piauí.

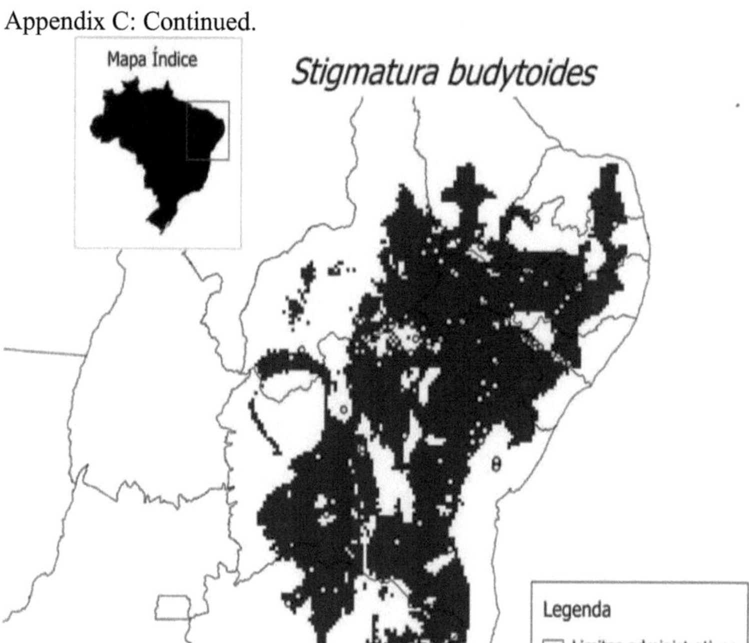

Stigmatura budytoides: It is distributed in the south-east of Piauí, south of Ceará and Paraíba, with a strip going up to the east of Rio Grande do Norte, the state of Pernambuco, except for the coast, west of Alagoas and Sergipe, going southwards covering the centre of Bahia and north of Minas Gerais, as well as the north of Espirito Santo.

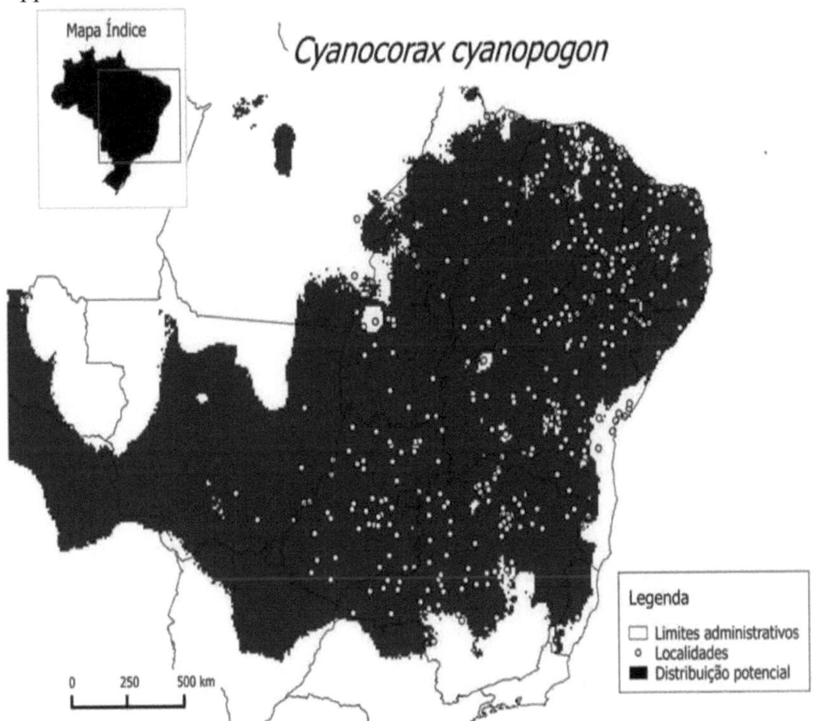

Cyanocorax cyanopogon: Occupies the states of Maranhão, Piauí, Ceará, Rio Grande do Norte, Paraíba, Pernambuco, Alagoas, Sergipe, Bahia, except for the coastal region, centre and west of Minas Gerais, north-west of Espirito Santo, as well as the states of Goiás, Tocantins and a strip of Mato Grosso, north of Mato Grosso do Sul and Rondônia.

Appendix C: Continued.

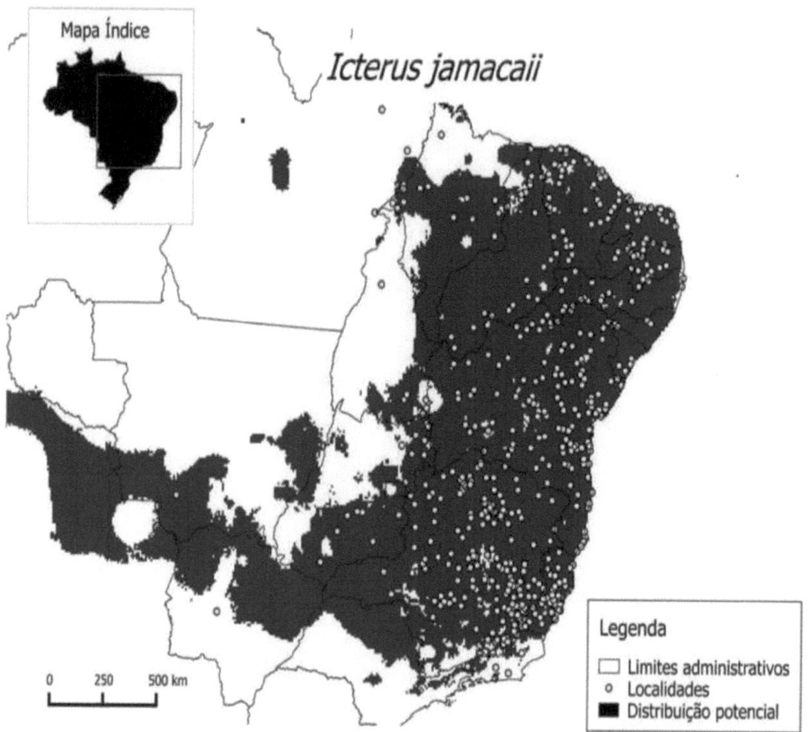

Icterus jamacaii: Distributed in the centre and south of Maranhão, throughout Piauí, Ceará, Rio Grande do Norte, Paraíba, Pernambuco, Alagoas, Sergipe, Bahia, Minas Gerais, Espirito Santo, north of Rio de Janeiro and São Paulo, east of Goiás, and a strip extending to the north of Mato Grosso do Sul and south of Mato Grosso.

Appendix C: Continued.

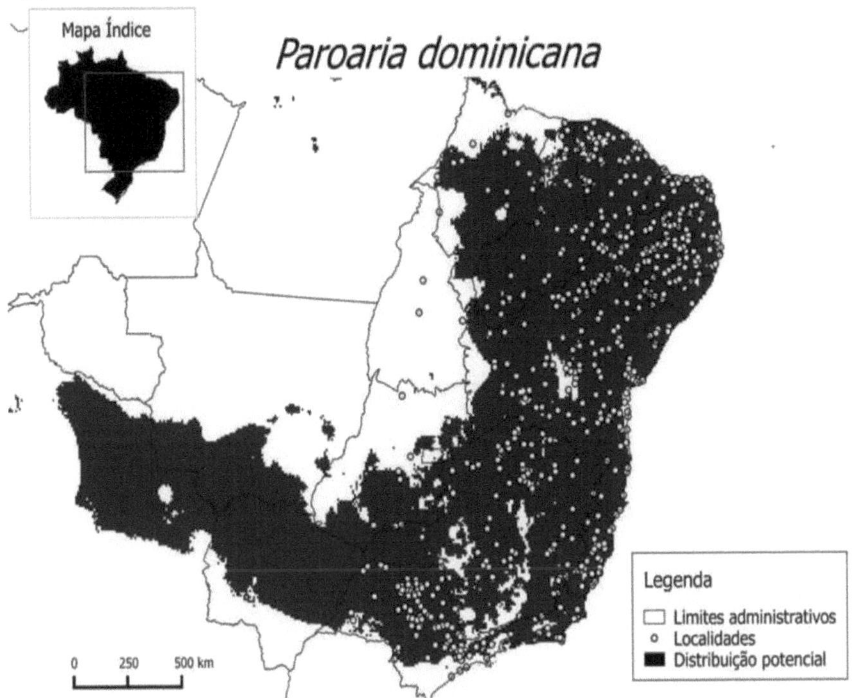

Paroaria dominicana: Distributed in south-central Maranhão, the entire state of Piauí, Ceará, Rio Grande do Norte, Paraíba, Pernambuco, Alagoas, Sergipe, Bahia, Minas Gerais, Espirito Santo, Rio de Janeiro, north-central São Paulo, south-eastern Goiás, and a strip extending to northern Mato Grosso do Sul and southern Mato Grosso.

Appendix C: Continued.

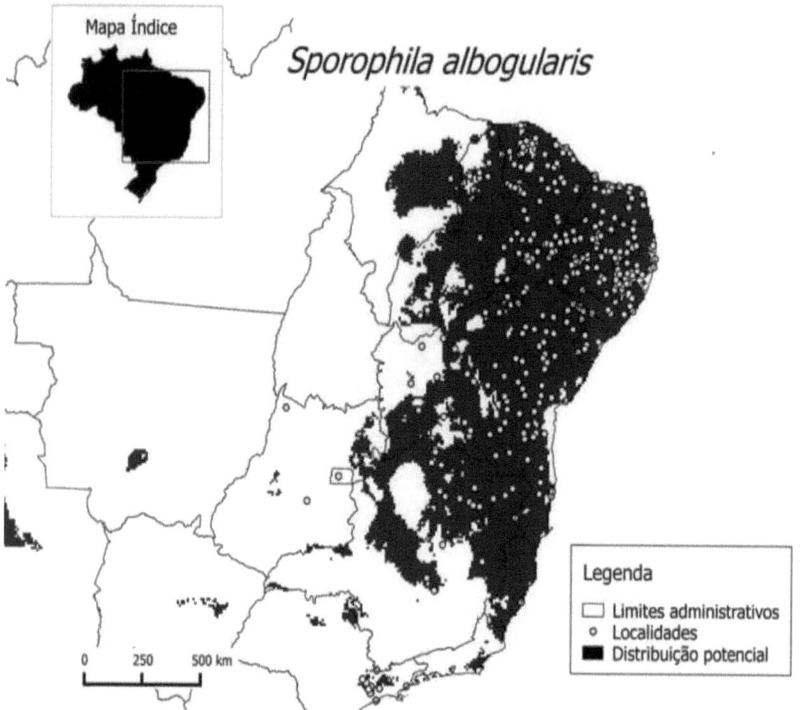

Sporophila albogularis: Distributed in north-eastern Brazil, including central Maranhão, the state of Piauí, Ceará, Rio Grande do Norte, Paraíba, Pernambuco, Alagoas, Sergipe, Bahia, northern Minas Gerais, Espirito Santo and a small patch on the coast of Rio de Janeiro and eastern São Paulo.

Appendix C: Continued.

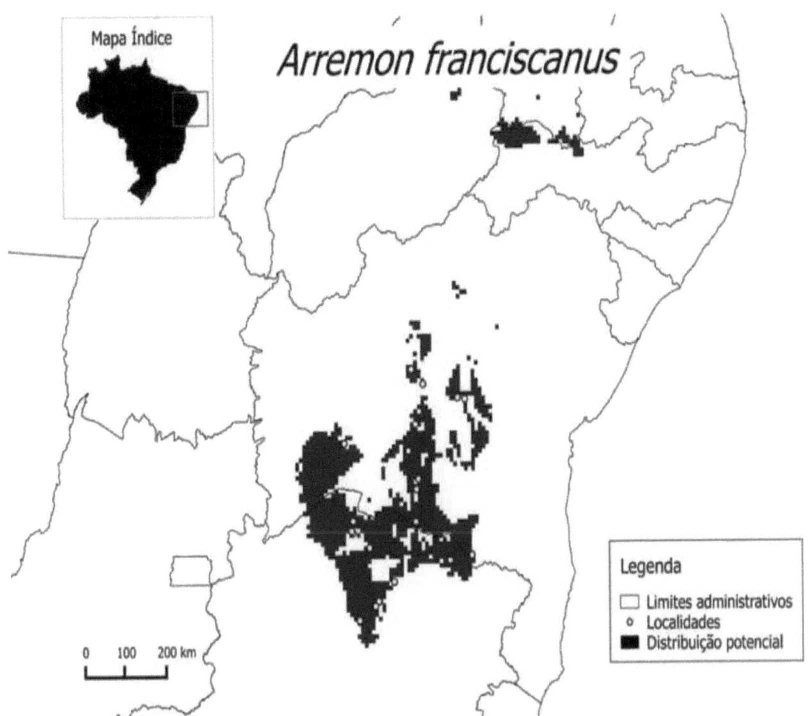

Arremon franciscanus: Distributed in the far north of Minas Gerais, northwards to the centre of Bahia.

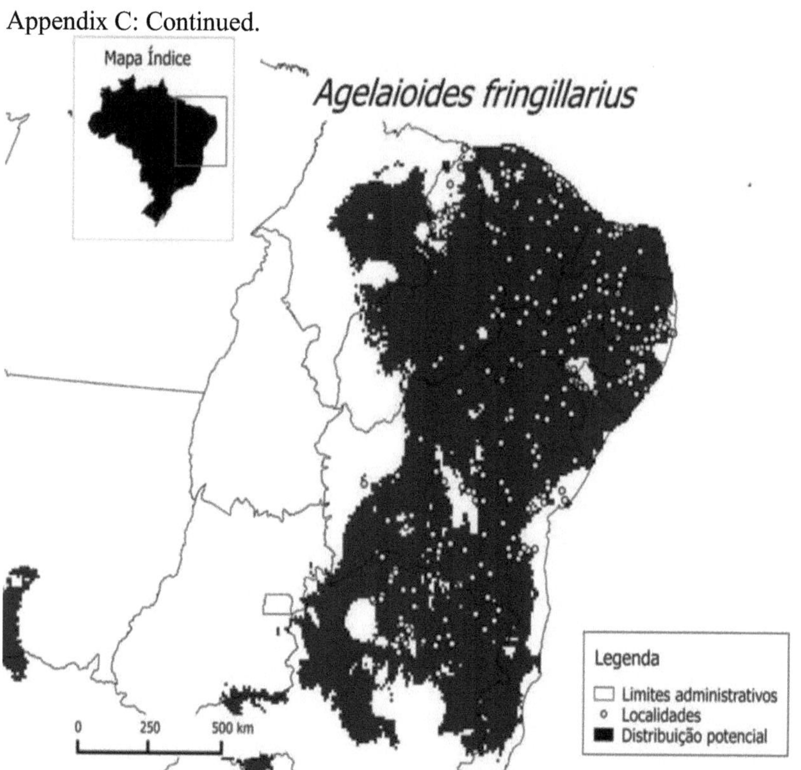

Agelaioides fringillarius: Occupies the central portion of Maranhão, heading east to Rio Grande do Norte, down to the north of Minas Gerais and Espirito Santo and occupying the central region of Bahia.

Appendix C: Continued.

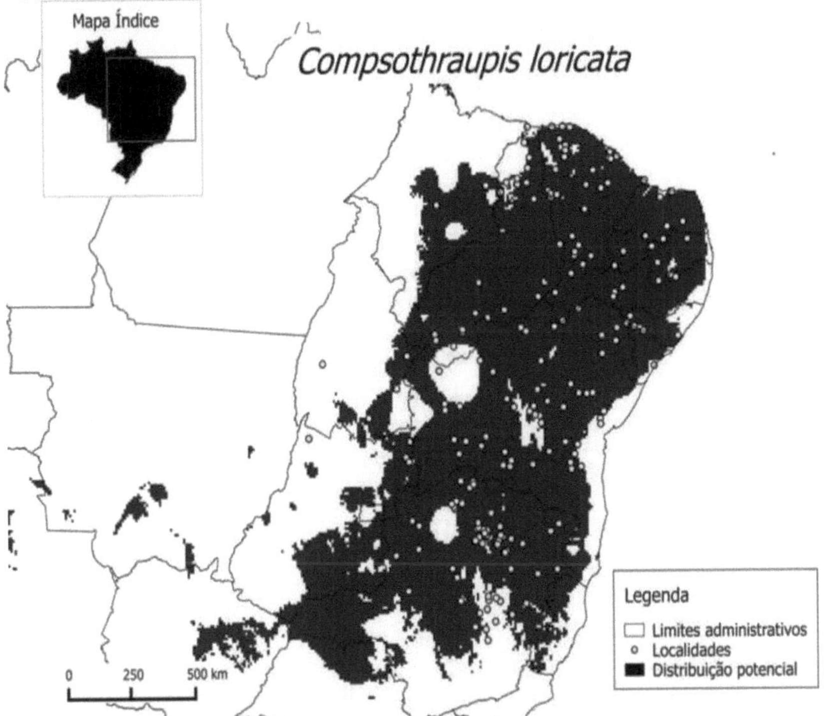

Compsothraups loricata: Distributed in north-eastern Brazil, except along the coast. From the south-east of Maranhão, towards Rio Grande do Norte, down through Bahia, to the centre-north of Minas Gerais, occupying a portion of eastern Goiás and western Espírito Santo.

Appendix D: List of the 309 municipalities that achieved suitability for most of the species modelled (over 19 species).

Municipality	State	Municipality	State
Abaiara	EC	Bodocó	PE
Abaré	BA	Bom Jardim	PE
Acauã	PI	Bom Jesus da Serra	BA
Acopiara	EC	Bonito de Santa Fe	PB
Afogados da Ingazeira	PE	Brejinho	PE
Afranio	PE	Brejo da Madre de Deus	PE
White Water	PB	Brejo Santo	EC
Água Nova	RN	Brumado	BA
Aguiar	PB	Cabrobó	PE
Aiuaba	EC	Indian Waterfall	PB
Alagoinha do Piauí	PI	Cachoeirinha	PE
Alegrete do Piauí	PI	Cacimba de Dentro	PB
Cotton from Jandaíra	PB	Cacimbinhas	AL
Altaneira	EC	Caculé	BA
Amparo	PB	Caetanos	BA
Anagé	BA	Cajazeiras	PB
Andaraí	BA	Caldeirão Grande do Piauí	PI
Antonina do Norte	EC	Campina Grande	PB
Araripe	EC	Campo Alegre do Fidalgo	PI
Araripina	PE	Campo Grande do Piauí	PI
Araruna	PB	Campo Redondo	RN
Areial	PB	Campos Sales	EC
Arneiroz	EC	Chapel of Alto Alegre	BA
Mastic trees	PB	Capim Grosso	BA
Assaré	EC	Captain Enéas	MG
Assunção do Piauí	PI	Caribbean	BA
Aurora	EC	Caridade do Piauí	PI
Avelino Lopes	PI	Caririaçu	EC
Baixio	EC	Cariús	EC
Barbalha	EC	Carnaíba	PE
Barra da Estiva	BA	Carnaubeira da Penha	PE
Barra de Santana	PB	Caruaru	PE
Clay	EC	Little houses	PE
Belém do Piauí	PI	Casserengue	PB
Belo Campo	BA	Catherine	EC
Belo Jardim	PE	Catuti	MG
Bernardino Batista	PB	Cedar	PE
Betânia do Piauí	PI	Central	BA

Boa Ventura	PB	Cerro Corá	RN
Boa Vista do Tupim	BA	Conceição	PB
Contendas do Sincorá	BA	Icó	EC
Coribe	BA	Igaracy	PB
Colonel Ezequiel	RN	Iguaraci	PE
Colonel João Pessoa	RN	Immaculate	PB
Colonel José Dias	PI	Inhuma	PI
Craíbas	AL	Ipaumirim	EC
Crato	EC	Ipubi	PE
Cuité	PB	Iramaia	BA
Curaçá	BA	Itaberaba	BA
Curimatá	PI	Itaeté	BA
Currais Novos	RN	Itaguaçu da Bahia	BA
Curral Novo do Piauí	PI	Itapetim	PE
Curral Velho	PB	Itaporanga	PB
Custody	PE	Itatuba	PB
Damião	PB	Itiúba	BA
Desterro	PB	Ituaçu	BA
Diamond	PB	Jaçanã	RN
Dom Basílio	BA	Jacobina do Piauí	PI
Dom Inocencio	PI	Jaguarari	BA
Dormant	PE	Jaíba	MG
Hope	PB	Jaicós	PI
Exu	PE	Janaúba	MG
Fagundes	PB	Japi	RN
Farias Brito	EC	Jaramataia	AL
Flowers	PE	Garden	EC
Francisco Macedo	PI	Jati	EC
Francisco Sá	MG	Jeremoabo	BA
Francisco Santos	PI	Joca Claudino	PB
Friar Michael	PE	José da Penha	RN
Borders	PI	Juazeiro	BA
Gado Bravo	PB	Juazeiro do Norte	EC
Game trees	MG	Jucás	EC
Hawkeye	BA	Juru	PB
Girau do Ponciano	AL	Jussara	BA
Granite	PE	Lagoa do Barro do Piauí	PI
Farmer	EC	Lagoa do Sítio	PI
Guajeru	BA	Royal Lagoon	BA
Ibiara	PB	Lagoa Seca	PB
Ibipeba	BA	Lajedinho	BA
Ibiquera	BA	Painted Slabs	RN
Lavras da Mangabeira	EC	Father Marcos	PI
Livramento	PB	Father Peter	MG

Livramento de Nossa Senhora	BA	Parnamirim	PE
Luís Gomes	RN	Patos do Piauí	PI
Maetinga	BA	Pau dos Ferros	RN
Major Isidoro	AL	Paulistana	PI
Malhada de Pedras	BA	Paulo Afonso	BA
Manaíra	PB	Pedra Branca	PB
Mango	MG	Pedro Alexandre	BA
Manoel Vitorino	BA	Penaforte	EC
Marcelino Vieira	RN	Petrolina	PE
Marcolândia	PI	Piancó	PB
Massapê do Piauí	PI	Picuí	PB
Massaranduba	PB	Pepper trees	PI
Matias Cardoso	MG	Painted	BA
Maturéia	PB	Pius IX	PI
Mauritius	EC	Pocinhos	PB
Miracles	EC	Dantas Well	PB
Minador do Negrão	AL	Gates	EC
Mirandiba	PE	Porteirinha	MG
Viewpoint	BA	Potengi	EC
Old Mission	EC	Silver	PB
Mombasa	EC	President Dutra	BA
Monsignor Hipólito	PI	President Jânio Quadros	BA
Mounted	PB	Princess Isabel	PB
Monte Azul	MG	Puxinanã	PB
Monte das Gameleiras	RN	Queimada Nova	PI
Mount Horeb	PB	Burning	PB
Monte Santo	BA	Rafael Fernandes	RN
Monteiro	PB	Remígio	PB
Moreilândia	PE	Riacho das Almas	PE
Morpará	BA	Riacho de Santana	RN
Nova Fátima	BA	Rio de Contas	BA
Nova Olinda	PB	Rio do Antônio	BA
New Redemption	BA	Ruy Barbosa	BA
Water eye	PB	Savoury	EC
Orobó	PE	Willow	PE
Orocó	PE	Saltpetre	EC
Ouricuri	PE	St Brigid	BA
Old Gold	PB	St Cecilia	PB
Santa Cruz	PE	Serra Grande	PB
Saint Philomena	PE	Serra Talhada	PE
St Helena	PB	Serrita	PE
Santa Inês	PB	Sertânia	PE
Santa Maria da Boa Vista	PE	Simões	PI

Santa Maria do Cambucá	PE	Sítio Novo	RN
Santa Terezinha	PE	Solânea	PB
Santaluz	BA	Sumé	PB
Santana de Mangueira	PB	Surubim	PE
Santana do Cariri	EC	Tabira	PE
Santana dos Garrotes	PB	Tacaimbó	PE
St Anthony of Lisbon	PI	Tangará	RN
São Bento do Trairí	RN	Tanhaçu	BA
São Bento do Una	PE	Taperoá	PB
São Caitano	PE	Taquaritinga do Norte	PE
São Domingos	BA	Tarrafas	EC
São Francisco de Assis do Piauí	PI	Tauá	EC
São Gabriel	BA	Tavares	PB
São João da Canabrava	PI	Terra Nova	PE
São João da Ponte	MG	Toritama	PE
São João das Missões	MG	Trinidad	PE
São José de Caiana	PB	Triumph	PB
São José de Piranhas	PB	Uauí	BA
St Joseph of Princesa	PB	Uibaí	BA
São José do Belmonte	PE	Uiraúna	PB
São José do Campestre	RN	Umari	EC
São José do Egito	PE	Umbuzeiro	PB
São José do Jacuípe	BA	Utinga	BA
São José do Piauí	PI	Valença do Piauí	PI
São José dos Cordeiros	PB	Valiant	BA
São Julião	PI	Várzea Alegre	EC
São Lourenço do Piauí	PI	Várzea da Roça	BA
São Luis do Piauí	PI	Come and see	RN
São Miguel	RN	Green	PE
São Miguel do Tapuio	PI	Verdelandia	MG
São Raimundo Nonato	PI	Lily Strand	PE
São Sebastião de Lagoa de Roça	PB	Strands	PE
São Sebastião do Umbuzeiro	PB	Vila Nova do Piauí	PI
St Thomas	RN	Wagner	BA
Sento Sé	BA	Xique-Xique	BA
Serra de São Bento	RN	Zabelê	PB

Appendix E: Areas proposed for conservation by The Nature Con- servancy of Brazil & Associação Caatinga (2000), Biodiversitas (2000), Velloso et al. (2001) and MMA (2002, 2004, 2007), which have had their importance corroborated. Legend: Adeq. =

suitability for x species; % = percentage of territory with high suitability.

Areas	Main town	State	Reference	Suita ble.	%
Araruna	Araruna	PB	Evaluation 2000; MMA 2004	20	100%
Sousa/ Valley of the Dinosaurs	Sousa	PB	Evaluation 2000; MMA 2004; MMA 2007	19 20	-< 50%
Serra da Joaninha / Serra da Pipoca	Quixeramobim, Tauá, Independência, Pedra Branca, Mombaça, Boa Viagem, Madalena, Canindé	EC	Biodiversitas 2000	19 22	-< 50%
Upper Piranhas Hinterland	São José da Lagoa Tapada, Coremas, São José de Piranhas	PB/ CE	Biodiversitas 2000	19 21	-< 50%
São José da Mata	Campina Grande, Po- cinhos, Puxinanã	PB	Biodiversitas 2000	19 20	-50%
Caruaru	Altinho, Agrestina, Bezerros, Cumaru, Caruaru	PE	Biodiversitas 2000	19 23	-> 50%
Buíque/ Ipojuca Valley	Capoeiras, São Bento do Una, Belo Jardim, Sanharó, Jataúba, Poção, Pesqueira	PE	Biodiversitas 2000	19 20	-< 50%
Curaçá	Santa Maria da Boa Vista, Orocó, Cabrobó, Curaçã, Juazeiro, Pe-trolina	PE/ BA	Biodiversitas 2000	19 20	-< 50%
Sento Sé	Sento Sé, Casa Nova	BA	Biodiversitas 2000	19 20	-< 50%
Ibotirama	Ibotirama	BA	Biodiversitas 2000	19 20	-100%
Ibipeba	Ibipeba	BA	Biodiversitas 2000	20 21	-100%

Carstê do Irecê	Lapão, América Dourada, Irecê, Juçara, Presidente Dutra, São Gabriel, Central, Uibaí, Ibititá, João Dourado	BA	Biodiversitas 2000	19 - 20	50%
Rui Barbosa	Rui Barbosa	BA	Biodiversitas 2000	19 - 21	< 50%
Livramento do Brumado	Dom Basílio, Érico Cardoso, Livramento de Nossa Senhora, Paramirim, Rio do Pires	BA	Biodiversitas 2000	19 - 22	> 50%
Aiuaba	Aiuaba, Saboeiro, Arneirós, Catarina	EC	Biodiversitas 2000	20 - 22	> 50%
Surroundings of Bom Jesus da Lapa	Bom Jesus da Lapa, Paratinga, Riacho de Santana	BA	Biodiversitas 2000	19 - 22	< 50%
Guanambi	Guanambi	BA	Biodiversitas 2000	20 - 21	< 50%
Peruaçu/ Jaíba	Jaíba, Itacarambi, Manga, Matias Cardoso, Januária, Monte Azul, Pedras de Maria da Cruz	MG	Biodiversitas 2000	19 - 21	50%
Serra Talhada	Serra Talhada	PE	Biodiversitas 2000, MMA 2004	20 - 21	100%
Raso da Catarina	Jeremoabo, Canudos, Glória, Macuru- ré, Santa Brígida, Paulo Afonso	BA	Biodiversitas 2000; MMA 2002	19 - 21	< 50%
Serra da Capivara	São João do Piauí, Coronel José Dias, São Raimundo Nonato, Canto	MA/ PI	Biodiversitas 2000; MMA 2002	19	< 50%

	do Buriti				
Maracás	Maracás, Jequié, Lafaiete Coutinho, Lajedo do Tabocal, Planaltino	BA	Biodiversitas 2000; MMA 2002	19 - 21	< 50%
Senhor do Bonfim	Senhor do Bonfim, Monte Santo, Uauá	BA	Biodiversitas 2000; MMA 2002	19 - 21	< 50%
Jaíba	Jaíba	MG	Biodiversitas 2000; MMA 2002	20 - 21	100%
Serra do Cariri	Imaculada, Água Branca, Juru, Princesa Isabel	PE/ PB	Biodiversitas 2000; MMA 2002, 2007	20 - 22	100%
Petrolina	Petrolina	PE	Biodiversitas 2000; MMA 2007	19 - 20	< 50%
Cariri Paraibano	-	PB	Biodiversitas 2000; Velloso 2001	19 - 23	< 50%
West of Pernambuco	Afrânio, Dormentes, Santa Cruz, Ouricuri, Parnamirim, Paulistana, Queimada Nova	PE/ PI	Biodiversitas 2000; Velloso 2001	19	> 50%
Delfino	Umburanas, Campo Formoso, Sento Sé	BA	Biodiversitas 2000; Velloso 2001	19 - 20	< 50%
Itaetê/ Abaíra	Andaraí, Boninal, Mucugê, Itaeté, Abaíra, Seabra, Piatã, Palmeiras, Lençóis, Ibicoara	BA	Biodiversitas 2000; Velloso 2001	20 - 22	< 50%
Peaks	Peaks	PI	Biodiversitas 2000; Velloso 2001; MMA 2002,2007	19 - 20	> 50%
Xingó	Paulo Afonso	BA/ AL/ SE	Biodiversitas 2000; Velloso	19 - 21	100%

				2001; MMA 2007	
Itacarambi/ Perua- çu	Jaíba, Itacarambi, Manga, Matias Cardoso, Januária, Monte Azul, Pedras de Maria da Cruz	MG	MMA 2002	19 - 21	> 50%
Picos/ Itainópolis	Picos, Itainópolis	PI	MMA 2002	19 - 20	> 50%
Two Brothers Mountain	Betânia do Piauí, Conceição do Canin- dé, Acauã, Jacobina do Piauí	PI/ PE	MMA 2002	19 - 20	< 50%
Surroundings of the Araripe National Forest	-	PI/ CE/ PE	MMA 2002	19 - 23	100%
Wonders	Custody	PE	MMA 2002	19 - 21	> 50%
Morpará/ Copixa- ba	Morpará, Xique- Xique	BA	MMA 2002	19 - 20	< 50%
Chapada Diamantina	-	BA	MMA 2002	19 - 23	< 50%
Janaúba	Janaúba	MG	MMA 2002	19 - 21	< 50%
Coremas	Coremas	PB	MMA 2002; MMA 2004	19	> 50%
APA São João do Piauí	São João do Piauí	PI	MMA 2004	19	< 50%
Exu National Park	Exu	PE	MMA 2004	19 - 23	> 50%
Araripina National Park	Araripina	PE	MMA 2004	19 - 21	100%
Source of the River Pardo	Jacaraci, Mortugaba, Montezuma, Espino- sa	BA/ MG	MMA 2004	19 - 22	50%
Southwest Bahia National Forest	Riacho de Santana, Matina	BA	MMA 2004	19 - 22	< 50%
Blue Macaw	Curaça, Juazeiro	BA	MMA 2004	19 - 20	< 50%
Casa Nova Ecological	Casa Nova, Petroli- na, Santa	BA/ PE	MMA 2004	19 - 20	> 50%

Station	Maria do Boa Vista, Afrânio				
Forest - Bank of the São Francisco	Petrolândia, Floresta, Tacaratu	PE	MMA 2004	19 - 20	< 50%
Martins National Park	Martins	RN	MMA 2004	19 - 20	< 50%
Luiz Gomes Biological Reserve	Luís Gomes	RN	MMA 2004	22	100%
Ceará/Piauí Conflict Area	-	CE/ PI	MMA 2004	19 - 20	< 50%
Dom Ino- cêncio APA	Dom Inocêncio, Lagoa do Barro do Piauí, Queimada Nova, Paulistana	PI	MMA 2004	19	> 50%
Serra do Barbado	Brumado	BA	MMA 2007	19 - 20	100%
Serra Brotas de Ma- caúbas	Xique-Xique	BA	MMA 2007	19 - 20	< 50%
Serra de Jacobina	Jacobina	BA	MMA 2007	19 - 22	> 50%
Gentio do Ouro	Xique-Xique	BA	MMA 2007	19 - 20	< 50%
Guigó from Coimbra	Our Lady of Glory	SE	MMA 2007	19	< 50%
Boqueirão	Petrolina	PE	MMA 2007	19 - 20	< 50%
Curaçá River and Mountains	Juazeiro	BA	MMA 2007	19 - 20	< 50%
São Francisco river channel	Petrolina	PE	MMA 2007	19 - 20	< 50%
Casa Nova	Petrolina	PE	MMA 2007	19 - 20	< 50%
Baixo da Melância	Santa Maria de Boa View	PE	MMA 2007	19	< 50%
Caboclo	Petrolina	PE	MMA 2007	19 - 20	< 50%
Serra do Arapuá	Serra Talhada	PE	MMA 2007	20 - 21	100%
Quilombola Community Conceição das Crioulas	of Willow	PE	MMA 2007	19 - 21	100%
Cabeceiras do Capibari- be	Caruaru	PE	MMA 2007	19 - 20	100%
Brejo Taquaritinga	de Caruaru	PE	MMA 2007	19 - 20	100%

Priority Areas	Main town	State	Reference		Suitable
Brejo da Princesa	Serra Talhada	PE	MMA 2007	20 - 21	100%
Conceição	Mauritius	EC	MMA 2007	22	100%
Chapada do Araripe (East)	Crato	EC	MMA 2007	21 - 22	< 50%
Fagundes	Campina Grande	PB	MMA 2007	19 - 20	> 50%
Piranhas	Cajazeiras	PB	MMA 2007	19 - 21	> 50%
Kariris	Juazeiro do Norte	EC	MMA 2007	20 - 21	100%
Cotton from Jandaíra	Hope	PB	MMA 2007	20 - 22	50%
Ink pear tree	Parambu	EC	MMA 2007	19 - 20	< 50%
Brejo do Cruz	Catolé do Rocha	PB	MMA 2007	19	> 50%
Santarém	Icó	EC	MMA 2007	20 - 22	< 50%
Pepper trees	São Miguel do Tapuio	PI	MMA 2007	19 - 21	> 50%
St Thomas	Currais Novos	RN	MMA 2007	19 - 21	50%
Castelo do Piauí	São Miguel do Tapuio	PI	MMA 2007	19 - 21	> 50%
Faveleira	Tauá	EC	MMA 2007	19 - 20	> 50%
Araripe Plateau	-	PE/ CE/ PI	MMA 2007; Biodiversity 2000	19 - 23	< 50%

Appendix F: Areas of medium and low richness proposed for conservation by The Nature Conser- vancy of Brazil & Associação Caatinga (2000), Biodiversitas (2000), Velloso et al. (2001) and MMA (2002, 2004, 2007), whose importance was corroborated due to the potential for endangered species.

Species	Priority Areas	Main town	State	Reference	Suitable.
Hemitric	Quixadá	Quixadá,	EC	Biodiversit	11 -

cus mirandae and Picumnus limae		Quixeramubim, Banabuiú, Choró		as (2000); Velloso (2001); MMA (2002)	17.
	Baturité Mountains	Aratuba, Capistrano, Itapiúna, Canindé, Caridade, Paramoti, Guaramiranga, Mulungu, Baturité, etc.	EC	Biodiversit as (2000); MMA (2004)	11 - 15.
	Northern Ibiapaba Plateau/ Jaburuna	Tianguá, Frecheirinha, Ubajara, Mucambo, Ibiapina, São Benedito, Graça, Carnaubal, etc.	EC	Biodiversit as (2000)	4 - 20.
	Pedra Branca	Quixeramobim	EC	MMA (2007)	16 - 17.
	São Joaquim	Quixeramobim	EC	MMA (2007)	16 - 17.
Hemitricus mirandae	Brejos de Natuba	Bom Jardim	PE	MMA (2007)	13 - 19.
Picumnus limae	Casta- nhão Ecological Station	Jaguaribara, Alto Santo	EC	Evaluation... (2000); MMA (2004)	15 - 17.
	Serra das Flores	Coreaú, Granja, Uruoca , Moraújo , Tiangua , Martinópole		Biodiversit as (2000)	10 - 17.
	Serra das Almas	Crateús, Conflict Zone, Buriti dos Montes	EC	Biodiversit as (2000)	13 - 19.
	Patos/ Santa Terezinha	Patos, Santa Teresinha	PB	Biodiversit as (2000)	13 - 15.
	São Bento do Norte	São Bento do Norte, Parazinho, Jandaíra	RN	Biodiversit as (2000)	15 - 18.
	Mato Grande	Lajes, Pedra Preta, Jandaíra, João Câmara, etc.	RN	Biodiversit as (2000)	12 - 18.
Picumnus limae	Acarí	Carnaúba dos Dantas, Acarí, Currais Novos, São Vicente	RN	Biodiversit as (2000); MMA (2002); MMA (2007)	13 - 18.
	Seridó/	Jardim de Piranhas,	RN/	Biodiversit	13 -

	Borborema	Brejo do Cruz, São Bento, Serra Negra do Norte, Paulista	PB	as (2000); Velloso (2001); MMA (2002)	15.
	Lower Jaguaribe/Chapada do Apodi	Aracati, Itaiçaba, Palhano, Russas, Ererê, Jaguaruana, Icapuí, Morada Nova	CE/ RN	Biodiversitas (2000); MMA (2004)	12 - 17.
	Morada Nova	Morada Nova	EC	MMA (2002)	13 - 17.
	Russas/ Icapuí	Russas, Icapuí	EC	MMA (2002)	13 - 16.
	Galinhos/ Jandaíra	Galinhos/ Jandaíra	EC	MMA (2002)	14 - 18.
	Crateús	Crateús	EC	MMA (2002); MMA (2007)	9 - 19.
	Jandaíra/ João Câmara	Jandaíra/ João Câmara	EC	MMA (2002); MMA (2007)	16 - 18.
	Tamanduá Farm	Santa Terezinha	PB	MMA (2002); MMA (2007)	13 - 15.
	Piracuruca	Piracuruca	PI	MMA (2004)	14 - 15.
Picumnus limae	São Bento do Norte National Park	Galinhos, Guamoré, São Bento do Norte, Pedra Grande, Touros	RN	MMA (2004)	11 - 18.
	Serra Caiada Natural Monument	Serra Caiada (former President Juscelino)	RN	MMA (2004)	18.
	Vale do Itaueira/ Gurgéia	Floriano	PI	MMA (2007)	12 - 17.
	Brejo	Guarabira	PB	MMA (2007)	17.
	Banana trees	Guarabira	PB	MMA (2007)	17.
	Mountain View	Paulista	PB	MMA (2007)	14 - 18.
	Forward	Aiuaba	EC	MMA (2007)	14 - 19.

	Tangará	Macaíba	RN	MMA (2007)	16 - 18.
	Martins	Pau dos Ferros	RN	MMA (2007)	16 - 18.
	Serra Micaela	Jaguaribe	EC	MMA (2007)	14 - 17.
	Caraúbas	Apodi	RN	MMA (2007)	11 - 17.
	Purity	Bulls	RN	MMA (2007)	16 - 18.
	Açu	Mossoró	RN	MMA (2007)	9 - 15.
	São Miguel	Bulls	RN	MMA (2007)	16 - 18.
	Tabuleiros of Caiçara do Norte	São Miguel do Gostoso	RN	MMA (2007)	15 - 17.
Picumnus limae	Serra do Estevão	Quixadá	EC	MMA (2007)	11 - 17.
	Mossoró River Estuary	Mossoró	RN	MMA (2007)	9 - 15.
	Rio Grande do Norte's internal platform	Bulls	RN	MMA (2007)	16 - 18.
	Icapuí	Aracati	EC	MMA (2007)	13 - 16.
	High Poty	Crateús	EC	MMA (2007)	9 - 19.
	Poranga	Crateús	EC	MMA (2007)	9 - 19.
	Coast Icapui/ Aracati	Aracati	EC	MMA (2007)	13 - 16.
	Serra do Machado/ Serra das Matas	Aracati	EC	MMA (2007)	13 - 16.
	Piranjí	Rattlesnake	EC	MMA (2007)	9 - 13.
	Aracolaba	Maranguape	EC	MMA (2007)	12 - 14.
	Beberibe	Rattlesnake	EC	MMA (2007)	9 - 13.
	Baturité Massif	Maranguape	EC	MMA (2007)	12 - 14.
	Bat Cave	Ipu	EC	MMA (2007)	13 - 18.

	Serra da Aratânia	Maracanáu	EC	MMA (2007)	10 - 11.
	Uruburetama mountain range	Itapipoca	EC	MMA (2007)	13 - 15.
	Mundaú	Itapipoca	EC	MMA (2007)	13 - 15.
	Paracuru/Trairi Reef Area	Paracuru	EC	MMA (2007)	10 - 13.
Picumnus limae	Itarema Estuarine Complex	Acaraú	EC	MMA (2007)	11 - 13.
	Timonha River Estuary	Barroquinha	EC	MMA (2007)	10.
	Maranguape mountain range	Caucaia	EC	MMA (2007); MMA (2004)	11 - 14.

Appendix G: Municipalities with a high degree of potential wealth that do not have Conservation Units or conservation actions.

Municipality	State	Municipality	State
Abaré	BA	Carnaubeira da Penha	PE
Acopiara	EC	Little houses	PE
Água Nova	RN	Casserengue	PB
Aiuaba	EC	Catherine	EC
Alagoinha do Piauí	PI	Catuti	MG
Cotton from Jandaíra	PB	Cerro Corá	RN
Amparo	PB	Coribe	BA
Anagé	BA	Colonel Ezequiel	RN
Antonina do Norte	EC	Colonel João Pessoa	RN
Areial	PB	Craíbas	AL
Arneiroz	EC	Cuité	PB
Assunção do Piauí	PI	Curimatá	PI
Avelino Lopes	PI	Curral Velho	PB
Baixio	EC	Damião	PB
Barra de Santana	BA	Desterro	PB
Clay	EC	Diamond	PB
Belém do Piauí	PI	Fagundes	PB
Belo Campo	BA	Francisco Macedo	PI
Bernardino Batista	PB	Francisco Sá	MG
Boa Ventura	PB	Francisco Santos	PI
Boa Vista do Tupim	BA	Gado Bravo	PB
Bom Jardim	PE	Game trees	MG
Bom Jesus da Serra	BA	Hawkeye	BA

Bonito de Santa Fe	PB	Girau do Ponciano	AL
Brejinho	PE	Farmer	EC
Indian Waterfall	PB	Guajeru	BA
Cachoeirinha	PE	Ibiara	PB
Cacimba de Dentro	PB	Igaracy	PB
Cacimbinhas	AL	Iguaraci	PE
Caculé	BA	Inhuma	PI
Caetanos	BA	Ipaumirim	EC
Campo Alegre do Fidalgo	PI	Iramaia	BA
Campo Grande do Piauí	PI	Itaguaçu da Bahia	BA
Campo Redondo	RN	Itapetim	PE
Chapel of Alto Alegre	BA	Itaporanga	PB
Capim Grosso	BA	Itatuba	PB
Captain Enéas	MG	Ituaçu	BA
Caribbean	BA	Jaçanã	RN
Caridade do Piauí	PI	Jaicós	PI
Cariús	EC	Japi	RN
Jaramataia	AL	Pepper trees	PI
Jati	EC	Painted	BA
Joca Claudino	PB	Pius IX	PI
José da Penha	RN	Dantas Well	PB
Jucás	EC	Porteirinha	MG
Lagoa do Sítio	PI	Silver	PB
Royal Lagoon	BA	President Jânio Quadros	BA
Lagoa Seca	PB	Burning	PB
Lajedinho	BA	Rafael Fernandes	RN
Painted Slabs	RN	Remígio	PB
Lavras da Mangabeira	EC	Rio do Antônio	BA
Livramento	PB	Savoury	EC
Maetinga	BA	St Cecilia	PB
Major Isidoro	AL	St Helena	PB
Malhada de Pedras	BA	Santa Inês	PB
Manaíra	PB	Santa Maria do Cambucá	PE
Manoel Vitorino	BA	Santaluz	BA
Marcelino Vieira	RN	Santana de Mangueira	PB
Massapê do Piauí	PI	Santana dos Garrotes	PB
Massaranduba	PB	St Anthony of Lisbon	PI
Maturéia	PB	São Bento do Trairí	RN
Minador do Negrão	AL	São Domingos	BA
Mirandiba	PE	São João da Canabrava	PI
Viewpoint	BA	São João da Ponte	MG
Monsignor Hipólito	PI	São José de Caiana	PB

Municipality	State	Municipality	State
Mounted	PB	St Joseph of Princesa	PB
Monte das Gameleiras	RN	São José do Belmonte	PE
Monteiro	PB	São José do Campestre	RN
Nova Fátima	BA	São José do Jacuípe	BA
Water eye	PB	São José do Piauí	PI
Orobó	PE	São Lourenço do Piauí	PI
Old Gold	PB	São Luis do Piauí	PI
Father Peter	MG	São Miguel	RN
Patos do Piauí	PI	São Sebastião de Lagoa de Roça	PB
Pau dos Ferros	RN	São Sebastião do Umbuzeiro	PB
Pedro Alexandre	BA	St Thomas	RN
Penaforte	EC	Serra de São Bento	RN
Piancó	PB	Serra Grande	PB
Picuí	PB	Sítio Novo	RN

Appendix G: Continued.

Municipality	State
Solânea	PB
Sumé	PB
Tacaimbó	PE
Tangará	RN
Taperoá	PB
Tarrafas	EC
Tavares	PB
Terra Nova	PE
Uauí	BA
Uiraúna	PB
Umari	EC
Umbuzeiro	PB
Valença do Piauí	PI
Valiant	BA
Várzea Alegre	EC
Várzea da Roça	BA
Come and see	RN
Green	PE
Verdelandia	MG
Lily Strand	PE
Vila Nova do Piauí	PI
Zabelê	PB

yes
I want morebooks!

Buy your books fast and straightforward online - at one of world's fastest growing online book stores! Environmentally sound due to Print-on-Demand technologies.

Buy your books online at
www.morebooks.shop

Kaufen Sie Ihre Bücher schnell und unkompliziert online – auf einer der am schnellsten wachsenden Buchhandelsplattformen weltweit! Dank Print-On-Demand umwelt- und ressourcenschonend produzi ert.

Bücher schneller online kaufen
www.morebooks.shop

Printed by Books on Demand GmbH, Norderstedt / Germany